The Overland Trail

Variables, Graphs, Linear Functions, and Equations

Teacher's Guide

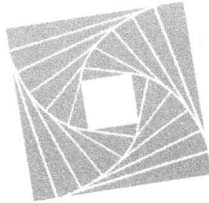

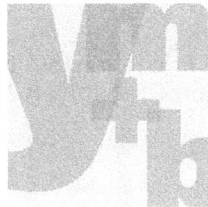

This material is based upon work supported by the National Science Foundation under award numbers ESI-9255262, ESI-0137805, and ESI-0627821. Any opinions, findings, and conclusions or recommendations expressed in this publication are those of the authors and do not necessarily reflect the views of the National Science Foundation.

Key Curriculum
1150 65th Street
Emeryville, California 94608
email: editorial@keypress.com
www.keycurriculum.com

First Edition Authors
Dan Fendel, Diane Resek, Lynne Alper, and Sherry Fraser

Contributors to the Second Edition
Sherry Fraser, Jean Klanica, Brian Lawler, Eric Robinson, Lew Romagnano, Rick Marks, Dan Brutlag, Alan Olds, Mike Bryant, Jeri P. Philbrick, Lori Green, Matt Bremer, Margaret DeArmond

Project Editors
Joan Lewis, Sharon Taylor

Consulting Editor
Mali Apple

Editorial Assistant
Juliana Tringali

Professional Reviewer
Rick Marks, Sonoma State University

Calculator Materials Editor
Christian Aviles-Scott

Math Checker
Christian Kearney

Production Director
Christine Osborne

Executive Editor
Josephine Noah

Textbook Product Manager
Timothy Pope

Publisher
Steven Rasmussen

Contents

Blackline Masters

Wagon Train Sketches and Situations Blackline Master
Graph Sketches Blackline Master
In Need of Numbers Blackline Master
The Issues Involved Blackline Master
Out Numbered Blackline Master
Broken Promises Blackline Master
1-Centimeter Graph Paper Blackline Master
1/4-Inch Graph Paper Blackline Master
1-Inch Graph Paper Blackline Master
In-Class Assessment
Take Home Assessment

Calculator Guide and Calculator Notes

Introduction

The Overland Trail Unit Overview

Intent

This unit uses the mid-nineteenth century migration of settlers from the eastern part of the United States across the west to California as the context for laying the foundation for algebraic thinking.

Mathematics

Building on students' work in *Patterns*, this unit develops the central mathematical idea of functions and their representations. Students will move among the following four "faces" of functions:

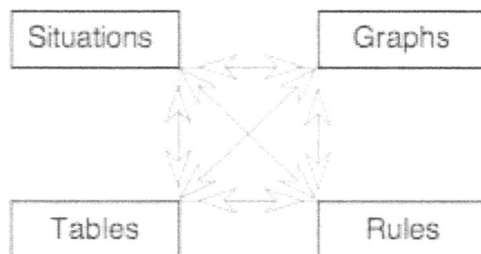

Situations	Graphs
Tables	Rules

The focus of this unit is on linear functions. Students will use starting values and rate of change to characterize linear functions, build In-Out tables, draw graphs, and write equations to represent specific contexts. They will use tables, graphs, and symbols to solve linear equations and systems of linear equations. They will fit lines to real data and use graphs and symbols representing these lines to solve problems in the context of the unit.

The main concepts and skills that students will encounter and practice during the course of this unit can be summarized by category.

Constraints and Decision Making

- Creating examples that fit a set of constraints
- Finding numbers that fit several conditions
- Using tables of information and lines of best fit to make predictions and estimates
- Working with mean and median

Algorithms, Variables, and Notation

- Strengthening understanding of the distributive property

- Developing numeric algorithms for problem situations
- Expressing algorithms in words and symbols
- Interpreting algebraic expressions in words using summary phrases
- Developing meaningful algebraic expressions

Basics of Graphing

- Reviewing the coordinate system
- Interpreting graphs intuitively and using graphs intuitively to represent situations
- Making graphs from tabular information
- Quantifying graphs with appropriate scales
- Using graphs to represent two-variable equations and data sets
- Using multiple representations—graphs, tables, and algebraic relationships—to describe situations

Linear Equations, Graphs, and Situations

- Finding and interpreting lines of best fit intuitively
- Seeing the role of constant rate in linear situations
- Using rates and starting values, or other data points, to create equations for straight lines
- Laying the groundwork for the concept of slope
- Using the point of intersection of two graphs to find values that satisfies two conditions
- Solving linear equations for one variable in terms of another
- Solving problems involving two linear conditions
- Solving linear equations in one variable

Graphs and Technology

- Making and interpreting graphs on a graphing calculator
- Using the zoom and trace features to get information from a graphing calculator

Progression

A wagon train trip from Missouri to California in the 1840s drives the unit. After students read about the history of westward migration and "adopt" families who are making the trip along the Overland Trail, they use variables and graphs to represent a variety of situations encountered by these families. They fit lines to data and identify the key features of linear relationships, and they develop graphing and symbol-manipulation skills to solve problems faced by travelers. Students also work on three POWs during the unit.

A Journey Back in Time

Setting Out with Variables

The Graph Tells a Story

Traveling at a Constant Rate

Reaching the Unknown

Supplemental Activities

Unit Assessments

Supporting Families Throughout *The Overland Trail*

Students do a series of activities that are directly related to the families they design in the beginning of the unit. The teaching suggestions in *Creating Families* ensure that students have resources available to do these activities.

In addition to work in *A Journey Back in Time,* the activities in the remainder of the unit bring forth opportunities to engage students with the Overland Trail families they identified with early on. What follows are ideas about preparing for and working with the families throughout *The Overland Trail*.

Organizing to Help Groups Create and Keep Track of Families

Teachers have found that they have more success with this aspect of the unit if they help students organize the record keeping for families, especially with regards to managing information about families of students who are absent so that other group members have the information they need.

Some teachers have designed handouts for students to fill in; others have encouraged students to be thoughtful about ways to organize information for themselves.

The activity itself requires students to record the name, age, and sex of each family member, for each of the group's four families. Students should also note the total number of adults and children in their group's families, as well as in the entire class wagon train, and keep this record available during the entire unit.

You can encourage students to be conscientious in creating these logs by sharing examples of similar logbooks kept in that era, or possibly from other migrations, such as logbooks kept of immigrants arriving at Ellis Island or slaves brought to trading ports in New Orleans.

You might use manila folders, one per group, in which students keep a copy of the information for each of their group's families. These folders can be kept in a convenient place.

Historical References

These books were used as source material in preparing this unit or have been used by teachers to supplement the historical information in the unit.

- Ralph K. Andrist, *The Long Death* (New York: Collier Books, Macmillan Publishing, 1964)

- Herbert Eaton, *The Overland Trail to California in 1852* (New York: Capricorn Books, G. P. Putnam's Sons, 1974)

- John Mack Faragher, *Women and Men on the Overland Trail* (New Haven: Yale University Press, 1979)

- Gwinn Harris Heap, *Central Route to the Pacific* (New York: Arno Press, 1981; reprint of the 1854 edition published by J. B. Lippincott Co.)

- James Hewitt, *Eye-Witnesses to Wagon Trains West* (New York: Charles Scribner's Sons, 1973)

- Kenneth Holmes, *Covered Wagon Women: Diaries and Letters from the Western Trails, 1840–1890*, vol. 1 (Glendale, CA: The Arthur H. Clark Co., 1983)

- William Loren Katz, *Black People Who Made the Old West* (Trenton, NJ: African World Press, 1992)

- Savoie Lotinville, *Life of George Bent* (Norman, OK: University of Oklahoma Press, 1968)

- Greg MacGregor, *Overland: The California Emigrant Trail of 1841–1870* (University of New Mexico Press, 1996)

- Peter Nabakov, ed., *Native American Testimony* (New York: Penguin Books, 1991)

- Lillian Schlissel, *Women's Diaries of the Westward Journey* (New York: Schocken Books, 1983)

- George R. Stewart, *The California Trail: An Epic with Many Heroes* (New York: McGraw-Hill Book Co., 1962)

You may also want to make use of one or more videos, such as these.

- *How the West Was Lost*, available from the Public Broadcasting System

- *The Donner Party*, available from the Public Broadcasting System

- *Gone West*, available from Social Studies School Service, 10200 Jefferson Blvd., Rm. 121, PO Box 802, Culver City, CA 90232-0802

Finally, many web sites are devoted to this topic. Here are three.

- http://www.tngenweb.org/tnletters/usa-west.htm

- http://freepages.genealogy.rootsweb.com/~gentutor/migration.html

- http://www.amaricanwest.com/pages/awexpans.htm

Pacing Guides

50-Minute Pacing Guide (30 days)

Day	Activity	In-Class Time Estimate
1	A Journey Back in Time	5
	Crossing the Frontier	40
	Homework: *Just Like Today*	5
2	Discussion: *Just Like Today*	10
	Overland Trail Families	10
	Creating Families	15
	POW 6: The Haybaler Problem	10
	Homework: *Hats for the Families*	5
3	Discussion: *Hats for the Families*	10
	Creating Families (continued)	30
	Homework: *Family Constraints*	10
4	Discussion: *Family Constraints*	20
	Planning for the Long Journey	25
	Homework: *The Search for Dry Trails*	5
5	Discussion: *The Search for Dry Trails*	15
	Setting Out with Variables	
	Shoelaces	25
	Homework: *Laced Travelers*	10
6	Discussion: *Laced Travelers*	15
	To Kearney by Equation	25
	Homework: *The Vermillion Crossing*	10
7	Discussion: *The Vermillion Crossing*	10
	Ox Expressions	30
	Homework: *Ox Expressions at Home*	10

8	Discussion: *Ox Expressions at Home*	15
	Presentations: *POW 6: The Haybaler Problem*	25
	Homework: *If I Could See This Thing*	10
9	Discussion: *If I Could See This Thing*	15
	The Graph Tells a Story	
	Wagon Train Sketches and Situations	30
	Homework: *Graph Sketches*	5
10	Discussion: *Graph Sketches*	15
	In Need of Numbers	30
	Homework: *The Issues Involved*	5
11	Discussion: *The Issues Involved*	20
	Out Numbered	25
	Homework: *From Rules to Graphs*	5
12	Discussion: *From Rules to Graphs*	20
	Out Numbered (continued)	15
	POW 7: Around the Horn	10
	Homework: *You're the Storyteller: From Rules to Situations*	5
13	Discussion: *You're the Storyteller: From Rules to Situations*	10
	Traveling at a Constant Rate	
	Previous Travelers	40
	Homework: *Broken Promises*	0
14	Discussion: *Broken Promises*	15
	Sublette's Cutoff	30
	Homework: *Who Will Make It?*	5
15	Discussion: *Who Will Make It?*	20
	The Basic Student Budget	25
	Homework: *Following Families on the Trail*	5
16	Discussion: *Following Families on the Trail*	20
	Graphing Calculator In-Outs	30
	Homework: *Fort Hall Businesses*	0

17	Discussion: *Fort Hall Businesses*	20
	"Sublette's Cutoff" Revisited	25
	Homework: *"The Basic Student Budget" Revisited*	5
18	Presentations: *POW 7: Around the Horn*	20
	Discussion: *"The Basic Student Budget" Revisited*	15
	POW 8: On Your Own	10
	Homework: *All Four, One*	5
19	Discussion: *All Four, One*	15
	Travel on the Trail	30
	Homework: *Moving Along*	5
20	Discussion: *Moving Along*	15
	All Four, One—Linear Equations	30
	Homework: *Straight-Line Reflections*	5
21	Discussion: *Straight-Line Reflections*	15
	Reaching the Unknown	
	Fair Share on Chores	30
	Homework: *Fair Share for Hired Hands*	5
22	Discussion: *Fair Share for Hired Hands*	15
	More Fair Share on Chores	30
	Homework: *More Fair Share for Hired Hands*	5
23	Discussion: *More Fair Share for Hired Hands*	15
	Water Conservation	35
	Homework: *The Big Buy*	0
24	Discussion: *The Big Buy*	15
	Discussion: *POW 8: On Your Own*	20
	The California Experience	10
	Homework: *Getting the Gold*	5
25	Discussion: *Getting the Gold*	15
	The Mystery Bags Game	30
	Homework: *More Mystery Bags*	5
26	Discussion: *More Mystery Bags*	20
	Scrambling Equations	25

	Homework: *More Scrambled Equations and Mystery Bags*	5
27	Discussion: *More Scrambled Equations and Mystery Bags*	20
	Family Comparisons by Algebra	25
	Homework: *Starting Over in California*	5
28	Discussion: *Starting Over in California*	20
	Beginning Portfolios	20
	Homework: *"The Overland Trail" Portfolio*	10
29	*In-Class Assessment*	50
	Homework: *Take-Home Assessment*	0
30	Discussion: *In-Class Assessment and Take-Home Assessment*	30
	Unit Reflection	20

90-Minute Pacing Guide (18 days)

Day	Activity	In-Class Time Estimate
1	A Journey Back in Time	
	Crossing the Frontier	35
	Just Like Today	20
	Overland Trail Families	10
	Creating Families	15
	POW 6: The Haybaler Problem	10
	Homework: *Hats for the Families*	0
2	Discussion: *Hats for the Families*	10
	Creating Families (continued)	30
	Family Constraints	45
	Homework: *The Search for Dry Trails*	5
3	Discussion: *The Search for Dry Trails*	20
	Planning for the Long Journey	30
	Setting Out with Variables	
	Shoelaces	30
	Homework: *Laced Travelers*	10
4	Discussion: *Laced Travelers*	15
	To Kearney by Equation	20
	The Vermillion Crossing	25
	Ox Expressions	20
	Homework: *Ox Expressions at Home*	10
5	Discussion: *Ox Expressions at Home*	10
	Presentations: *POW 6: The Haybaler Problem*	25
	If I Could See This Thing	25
	The Graph Tells a Story	
	Wagon Train Sketches and Situations	25
	Homework: *Graph Sketches*	5

6	Discussion: *Graph Sketches*	15
	In Need of Numbers	25
	The Issues Involved	35
	Out Numbered	15
	Homework: *From Rules to Graphs*	0
7	Discussion: *From Rules to Graphs*	20
	Out Numbered (continued)	25
	You're the Storyteller: From Rules to Situations	30
	POW 7: Around the Horn	15
8	Traveling at a Constant Rate	
	Previous Travelers	35
	Broken Promises	25
	Sublette's Cutoff	25
	Homework: *Who Will Make It?*	5
9	Discussion: *Who Will Make It?*	15
	The Basic Student Budget	20
	Following Families on the Trail	35
	Graphing Calculator In-Outs	20
	Homework: *Fort Hall Businesses*	0
10	Discussion: *Fort Hall Businesses*	20
	"Sublette's Cutoff" Revisited	25
	"The Basic Student Budget" Revisited	45
	Homework: *All Four, One*	0
11	Discussion: *All Four, One*	20
	Presentations: *POW 7: Around the Horn*	20
	POW 8: On Your Own	15
	Travel on the Trail	30
	Homework: *Moving Along*	5
12	Discussion: *Moving Along*	10
	All Four, One—Linear Equations	25
	Straight-Line Reflections	25

	Reaching the Unknown	
	Fair Share on Chores	25
	Homework: *Fair Share for Hired Hands*	5
13	Discussion: *Fair Share for Hired Hands*	10
	More Fair Share on Chores	25
	More Fair Share for Hired Hands	30
	Water Conservation	25
	Homework: *The Big Buy*	0
14	Discussion: *The Big Buy*	10
	Discussion: *POW 8: On Your Own*	15
	The California Experience	10
	Getting the Gold	30
	The Mystery Bags Game	25
	Homework: *More Mystery Bags*	0
15	Discussion: *More Mystery Bags*	15
	Scrambling Equations	20
	More Scrambled Equations and Mystery Bags	30
	Family Comparisons by Algebra	25
	Homework: *Starting Over in California*	0
16	Discussion: *Starting Over in California*	30
	Beginning Portfolios	40
	Homework: *"The Overland Trail" Portfolio*	20
17	*"The Overland Trail" Portfolio*	30
	In-Class Assessment	50
	Homework: *Take-Home Assessment*	10
18	Discussion: *In-Class Assessment and Take-Home Assessment*	50
	Unit Reflection	40

Materials and Supplies

All IMP classrooms should have a set of standard supplies and equipment, and students are expected to have materials available for working at home on assignments and at school for classroom work. Lists of these standard supplies are included in the section "Materials and Supplies for the IMP Classroom" in *A Guide to IMP*. There is also a comprehensive list of materials for all units in Year 1.

Listed below are the supplies needed for this unit. General and activity-specific blackline masters are available for presentations on the overhead projector or for student worksheets. The masters are found in the *The Overland Trail* Unit Resources under Blackline Masters.

The Overland Trail

- Dice (at least one pair per group)
- Wall-size map of the United States
- Clear straightedges (rulers) or uncooked spaghetti for finding the line of best fit
- Colored pencils
- File folders for groups to keep information about their Overland Trail families
- Sentence strips for posting results
- Index cards

More About Supplies

- Sentence strips are useful in many IMP units. They are often used for posting solutions to problems, for posing problems, and for posting comments, strategies, and questions. These strips can be purchased at educational supply stores or simply made by cutting strips of construction paper, butcher paper, or chart paper.

- A wall size map of the United States helps students follow the progress of their Overland Trail families throughout the unit. A wall size map of the actual Overland Trail was published in the September 2000 edition of the National Geographic magazine. A smaller version of this map and a wonderful resource article, "The Way West," are available on the National Geographic Web site at http://www.nationalgeographic.com/ngm/0009/feature2/index.html.

- Graph paper is a standard supply for IMP classrooms. Blackline masters of 1-Centimeter Graph Paper, 1/4-Inch Graph Paper, and 1-Inch Graph Paper are provided so you can make copies and transparencies for your classroom. (You'll find links to these masters in "Materials and Supplies for Year 1" of the Year 1 guide and in the Unit Resources for each unit.)

Assessing Progress

The Overland Trail concludes with two formal unit assessments. In addition, there are many opportunities for more informal, ongoing assessment throughout the unit. For more information about assessment and grading, including general information about the end-of-unit assessments and how to use them, see "Assessment and Grading" in *A Guide to IMP*.

End-of-Unit Assessments

Each unit concludes with in-class and take-home assessments. The in-class assessment is intentionally short so that time pressures will not affect student performance. Students may use graphing calculators and their notes from previous work when they take the assessments. You can download unit assessments from *The Overland Trail* Unit Resources.

Ongoing Assessment

Assessment is a component in providing the best possible ongoing instructional program for students. Ongoing assessment includes the daily work of determining how well students understand key ideas and what level of achievement they have attained in acquiring key skills.

Students' written and oral work provides many opportunities for teachers to gather this information. Here are some recommendations of written assignments and oral presentations to monitor especially carefully that will offer insight into student progress.

- *Creating Families:* This assignment will give you information on how well students can deal with verbal constraints.

- *Laced Travelers:* This activity will tell you whether students can put arithmetic processes into words.

- *Ox Expressions at Home:* This assignment will help you assess how well students understand meaningful algebraic expressions

- *Graph Sketches:* This activity will give you a sense of how well students understand graphs.

- *Who Will Make It?* This activity can help you gauge students' ability to make meaningful inferences from graphs.

- *All Four, One—Linear Functions:* This assignment will give you information about students' understanding of the connections among different ways to represent a situation.

- *Straight-Line Reflections:* This activity will give you a sense of how well students understand concepts related to straight-line graphs.

- *More Fair Share for Hired Hands:* This assignment can provide information on student understanding of the connection between graphs and equations.

- *Family Comparisons by Algebra:* This activity will help you evaluate students' ability to represent situations using equations and their facility with solving linear equations.

Supplemental Activities Overview

The Overland Trail contains a variety of activities at the end of the student pages that you can use to supplement the regular unit material. These activities fall roughly into two categories.

- **Reinforcements** increase students' understanding of and comfort with concepts, techniques, and methods that are discussed in class and are central to the unit.

- **Extensions** allow students to explore ideas beyond those presented in the unit, including generalizations and abstractions of ideas.

The supplemental activities are presented in the teacher's guide and the student book in the approximate sequence in which you might use them. Below are specific recommendations about how each activity might work within the unit. You may wish to use some of these activities, especially the later ones, after the unit is completed.

Pick Any Answer (reinforcement) This activity, which brings out the importance of the sequence in which arithmetic operations are done, provides a simple context for which students can create an algebraic expression. It might be an appropriate activity to use in connection with students' work with variables.

Substitute, Substitute (reinforcement) This activity provides examples through which students can strengthen their understanding of the basic ideas of substitution. The activity uses a variety of phrases for referring to the process. One question uses substitution to reinforce the idea of combining terms.

From Numbers to Symbols and Back Again (extension) This activity uses formulas from two settings from earlier units (*The Game of Pig* and *Patterns*) as the context for work with substitution. Students will have to guess and test solutions to Questions 1c and 2b, as the equations are quadratic. Furthermore, the solution to Question 1c is not integral (or even rational), so students will have to determine a reasonable approximation.

Classroom Expressions (reinforcement) Students create summary phrases and algebraic expressions using a set of variables that relate to a classroom setting. The activity also introduces the notation of subscripts.

Variables of Your Own (reinforcement) Students make up a set of variables, and write algebraic expressions and summary phrases, in a context of their own choosing.

Painting the General Cube (extension) In this activity, students create equations to describe a geometric situation.

Integers Only (extension) The activity *If I Could See This Thing* involves a function with outputs that should only be integers. *Integers Only introduces* the greatest integer function. Interested students might try to use this function to devise a formula for Question 2c of *If I Could See This Thing*.

More Bales of Hay (extension) This activity makes a good follow-up to presentations of *POW 6: The Haybaler Problem.* Experimentation with numbers is the essence of this activity. In particular, students are asked to consider whether such problems always have unique answers and in which cases the answers are whole numbers.

Spilling the Beans (reinforcement or extension) An essential part of this activity is the need to clearly understand any assumptions one makes. The activity also involves proportional reasoning, which plays an important role in the last unit of Year 1, *Shadows*.

More Graph Sketches (reinforcement) This activity provides a variety of contexts for which students can create graph sketches like those in *Graph Sketches*.

Movin' West (extension) This activity deals with questions about migration patterns in the United States and whether one can expect such patterns to continue, so it is a suitable thematic follow-up to *Broken Promises*. In addition to developing a general algebraic formula for a rate-of-change situation, students must think about comparative rates of change and do some commonsense reasoning to decide how well the model suggested by their formula might work in the future.

What We Needed (reinforcement) This activity asks students to figure out how long it took for their families to travel from Ft. Laramie to Ft. Hall and how much of two commodities they would need to bring with them.

Mystery Graph (reinforcement) In this graph-interpretation activity, students are given the graph of a nonlinear function, but not its equation, and are asked to find a number of values for the function.

POW: High Low Differences This activity is an additional open-ended POW in which students investigate sequences of calculations and look for patterns.

High Low Proofs (extension) This activity asks students to prove conjectures they made in the supplemental *POW: High Low Differences* explorations.

Keeping Track, A Special Show, and *Keeping Track of Sugar* (reinforcement) These activities offer additional opportunities for students to work with situations involving constant rates of change. All three involve constructing and using equations given two data points. You might use them early in *Traveling at a Constant Rate* if students are having difficulty creating equations to describe such situations.

The Growth of Westville (extension) The work early in *Traveling at a Constant Rate* focuses on situations involving constant rates of change and that led to linear graphs. This activity provides a western setting for examining situations that may appear to involve constant growth but do not lead to linear graphs, so this is a good follow-up to the series of activities.

Westville Formulas (extension) This activity is a follow-up to the supplemental activity *The Growth of Westville.*

The Perils of Pauline (extension) This is a well-known but challenging puzzle problem. Students are given information about the speed of an oncoming train and the position of a person in a tunnel that the train is approaching and are asked to determine the person's speed given that she made it out of the tunnel on time.

A Journey Back in Time

Intent

The activities in *A Journey Back in Time* are designed to accomplish two goals: to orient students to the historical context of the unit and to begin the development of one of the key mathematical ideas that drives the unit.

Mathematics

This unit develops the mathematical idea of variables and explores equations and graphs of linear functions. In *A Journey Back in Time*, students will begin to use variables and equations in a *representational* way—to stand for variable quantities and the relationships among them—as the precursor for being able to understand the algebraic manipulations that form the core of most school algebra programs but are often taught in purely procedural terms.

Progression

The activities in *A Journey Back in Time* set the historic context of the unit. The introductory activities offer students a sense of the history surrounding the Overland Trail, which is a collective name for a group of trails that led from Missouri to the West. In the last few activities, students encounter the mathematics of algebraic representation and equation solving. In addition, they begin work on the first POW of the unit.

Crossing the Frontier

Just Like Today

Overland Trail Families

Creating Families

Overland Trail Names

POW 6: The Haybaler Problem

Hats for the Families

Family Constraints

Planning for the Long Journey

Overland Trail Price List

The Search for Dry Trails

Crossing the Frontier

Intent

This reference page will draw students into the historic setting of the unit.

Mathematics

Although no particular mathematics concepts are addressed, this page introduces a historic context with which students will develop a personal connection, motivating their commitment to engaging in the activities to follow.

Progression

Students read the passage in the student book. This can be followed by further research about the time period and an opportunity to reflect through a focused free-write. *Just Like Today* is a good follow-up homework activity.

The geographical information in this unit is accurate to the best of our knowledge. The times given for various sections of the journey are reasonable estimates based on sources from and about the period. Information on prices from the period is based on sources where so noted, but otherwise may not reflect actual values at the time.

Approximate Time

40 minutes

Classroom Organization

Whole group, followed by further study in pairs or small groups, and then individuals for a focused free-write.

Materials

Large U.S. map for indicating the route of and landmarks along the Overland Trail

Additional resources to engage students in the historic and geographic context of the Overland Trail, such as video, pictures, books, and Internet resources, can be quite helpful. See the list of references in the overview to this unit.

Using the Reference Page

If possible, display a large wall map and help students trace the route of the California Trail, which began in Westport, Missouri (near present-day Kansas City), and ended at Sutter's Fort, California. Most of this trail was originally known as the Emigrant Trail, or Oregon Trail, which ended near what is now Portland, Oregon. The California Trail split from the Oregon Trail west of Fort Hall, which was located in what is now Pocatello, Idaho. Identify the modern-day states that the trail passes through: Missouri, Kansas, Nebraska, Wyoming, Idaho, Nevada, and California. Students generally find personal connections to the map and landmarks along the way, researched possibly via the Internet, very stimulating.

Identify the Native American nations through whose lands the trail ran. The trail began in Shawnee, Kansas, Lenni-Lenape (Delaware), Cheyenne, and Arapaho territory. The Dakota nation held the land that is now Nebraska and part of Wyoming. The Shoshone, Assiniboin, and Crow nations held the land of present-day Wyoming and Idaho. The Oregon Trail cut through Choppunish (Nez Percé), Flathead, Yakima, and Chinook territory. Some 300 nations, including the Modoc, Washo, Maidu, Pomo, and Miwok, held territory in what is now California.

You might mark these Native American nations on the map, where the trail traverses their lands.

By the mid-1800s, the U.S. government had appropriated Native American lands in the East and Midwest and moved whole Native American nations westward. Prior to 1840, the government "reserved" the land known as the Great Plains for those displaced peoples, calling it Indian Country. The established boundary of Indian Country started in the South at the edge of the Republic of Texas and ran north along the western boundaries of Arkansas and Missouri, across the northern border of Missouri roughly to the Mississippi River. It followed the Mississippi into Minnesota before cutting across Wisconsin on an irregular line to Lake Michigan. The map for 1830 in *Broken Promises* shows approximately this boundary.

The land was pledged to Native Americans to have and to hold forever. The U.S. government liked to use the phrase "as long as waters flow and the grass shall grow" in its treaties. This land was to be the final home for the displaced tribes from the East and for the existing tribes of the plain. Between 1778 and 1868, Native American nations and the U.S. government signed more than 400 treaties and agreements. The U.S. government broke every one of them. The government record of broken treaties is referred to in *Broken Promises*.

There are references in the unit to the devastating effects of western migration on Native Americans, but this devastation cannot be overstated. Disease alone reduced the native population of North America by over 80 percent. Any stories that you know or find will no doubt add more depth to the unit.

Students will be more engaged in the unit if they can make a connection with people who made the journey. This connection can be strengthened by looking at video depicting the era, exploring Web resources, and other research work, either now or at a later point during the unit. You might also bring in photographs and drawings of the era so that students have visual images to relate to.

When the following names appear in the activities, let students know they are reading about people who actually lived during this era. Other characters in the activities are fictitious.

- Joseph and Louis Papan (*To Kearny by Equation*)
- Louis Vieux (*The Vermillion Crossing*)
- George Bent (*If I Could See This Thing*)
- James P. Beckwourth (*Travel on the Trail*)
- Biddy Mason (*Starting Over in California*)

After introducing the unit using the student page *A Journey Back in Time*, use *Crossing the Frontier* to give a brief picture of the circumstances that led to the westward migration and of the journey itself. You might ask volunteers to read aloud as the class follows along.

To raise curiosity about the era, provide some time for discussion and for sharing information, pictures, maps, or other resources you have gathered. You might also design student information-gathering teams and encourage sharing of their findings.

After students view a video or read background material, you might have them do focused free-writing in preparation for a class discussion. Give students a full 5 minutes to write. Here are a few topics they might write about.

What did you learn from the video about the Overland Trail?

What do you know about this period of time in U.S. history?

If you traveled back in time to that period, what would you want to bring with you?

You might ask volunteers to share from their writing as a way to summarize or clarify ideas.

Just Like Today

Intent

In this continued introduction to the unit, students consider the why and how of movements of people in more recent times.

Mathematics

No specific mathematics concepts are addressed in this introductory writing activity to the historic context of the unit.

Progression

After some reflective writing time, students share ideas with classmates.

Approximate Time

5 minutes for introduction

15 minutes for activity (at home or in class)

10 minutes for discussion

Classroom Organization

Individuals, followed by small-group sharing of responses

Doing the Activity

To introduce this writing activity, draw upon the discussion of *Crossing the Frontier* to suggest that there are ways in which groups of people today have experiences similar to those of European settlers traveling along the Overland Trail during the mid-1800s.

Discussing and Debriefing the Activity

Have students work in their groups to share the comparisons they made of two movements. You may also want to ask for a few volunteers to share their ideas with the class.

Point out that while the events the class will be studying happened about 150 years ago, many of the issues that were important then are still important today.

Key Question

How did the movement you wrote about compare to the Overland Trail movement?

Overland Trail Families

Intent

Students read this reference page about extended family units on the Overland Trail in preparation for constructing families of their own.

Mathematics

This information continues to develop the context in which students will construct, interpret, and manipulate algebraic expressions and equations.

Progression

Students read this diary entry in preparation for the activity *Creating Families*.

Approximate Time

10 minutes

Using the Reference Page

Tell students that, during this unit, they will be planning and carrying out a trip of their own along the Overland Trail. The first step will be to develop a family of travelers for which to be responsible.

Have students read the excerpt from the diary of Catherine Haun aloud to learn more about families that traveled the trail. Spend some time, either during the reading or afterward, helping students develop an image of what is described in the diary. There will be terms many students may be unfamiliar with, such as *schooner, oxen,* and *draught.*

Creating Families

Intent

Beyond creating a more personalized experience of the mathematics in the unit, this activity engages students in the mathematical task of meeting conditions in order to determine possible solutions for a problem situation.

Mathematics

Students assemble groups of people that match a set of constraints derived from possible scenarios of westward migration. Although the methods used are informal, this serves as an initial experience in interpreting mathematically imprecise language. When combined with its companion activity, *Hats for the Family,* the activity gives students experience in the mathematical task of finding minimum and maximum numbers based on their interpretation of the set of constraints.

Progression

Students work in groups to create four hypothetical family units, information that will be referred to throughout the unit.

Approximate Time

45 minutes

Classroom Organization

Groups

Materials

Folders (1 per group)

Doing the Activity

The activity describes four types of family units that might have been part of a wagon train. Each group of students will create one family of each type. During the course of the unit, groups will follow the planning and movement of these families along the trail. Keeping track of these families helps to make the unit's context more real for students. "Supporting Families Throughout the Overland Trail" offers suggestions for helping groups create and keep track of families.

Groups will name their family members by drawing from the list of settlers' names in *Overland Trail Names.* Some of these names will appear in later activities. Seeing "their" family members in these situations will enhance students' engagement in the unit.

Emphasize that each group as a whole is responsible for creating the families and meeting the given constraints, even though each student will have final responsibility for one particular family.

Adjust accordingly for groups with other than four members. It is essential that each group create and keep track of one of each type of family. You might have some students keep track of more than one family or have the entire group monitor an unassigned family.

While students work, observe and support group interaction and careful record keeping. You or groups must decide how to record the information about each group's four families. Groups might keep their information in a folder while the class creates a chart summarizing the information about all families the class has created.

As you observe groups working, you may find it useful to have a sense of the size and structure of each family type. Here is a summary of some of the limitations that result from specific interpretations of the descriptions. For instance, "between one and six hired hands" is interpreted to mean "at least one and at most six." Students might reach slightly different conclusions, based on different interpretations. Students will determine maximum and minimum family sizes in *Hats for Families*.

- *Small family:* This family will have at least 3 and at most 7 people, with the smallest possible family consisting of a pair of adult siblings and one child.

- *Large family:* Assuming children are not married and "between" is interpreted as inclusive of 1 and 6, this family will have at least 13 and at most 25 people, with the smallest possible family consisting of 6 adults and 7 children. The 6 adults could be a married couple (the parents), the married parents of one of them (the grandparents), one great-grandparent, and one hired hand.

- *Nonfamily:* This group will have at least 2 and at most 12 people, with the smallest possible family consisting of two male adults.

- *Conglomerate family:* This family will have at least 3 and at most 12 people, with the smallest possible family consisting of two partial families, one with just one adult, the other with an adult and a young child.

Altogether, a wagon train consisting of one family of each type will have from 21 to 56 people.

It can be useful to bring the class together about one third of the way through the activity to check in. Ask what has been challenging. Replies may include monitoring group processes, record keeping, and meeting the demands of the family conditions.

There are some ambiguities in the wording of the constraints, such as use of the word *between* in the description of the large family. If students raise this issue, urge them to make decisions as a group about how to interpret these ambiguities.

Confirm the manner in which students are expected to record final decisions about the families: first and last name, age, and sex. Each student must have access to her or his own data, and each group member's data, at all times.

As groups near the finish of their work, ensure that they have met the requirements for recording the information about each group member's family. Depending on how you have decided to support student record keeping (perhaps using individual file folders or charts), remind students to complete that task as well.

Discussing and Debriefing the Activity

When all groups are done creating their families, compile a class chart summarizing information about all of the families created. Have presenters talk about any assumptions they made. Organize the information in a chart like the one below or in another way that students suggest.

	Group Number							
	1	2	3	4	5	6	7	8
Small family								
Large family	[Entries would show the number of men,							
Nonfamily	women, and children in each family.]							
Conglomerate family								

Post this chart for reference throughout the unit. Point out that some families may require more than one wagon; students will use this fact in later activities. Also for use later, ask students to find the total number of adults and the total number of children in their entire class wagon train.

Make sure students monitor the progress of their Overland Trail families throughout the unit. To do so, they will need to keep track of certain information. Specifically, *each student* will need to know:

- The number of men, women, and children in each of the four families created by the student's own group
- Which family the student is personally responsible for
- The total number of adults and the total number of children in the entire wagon train, which consists of all the individual families created by all the groups

Reading this list aloud can give students one last opportunity to double check that they have each of these pieces of information recorded.

Key Questions

What has been challenging for your group in this activity?

How did your group decide to interpret that constraint?

What assumptions did you make?

How many men, women, and children are in each of your families?

Overland Trail Names

Intent

Students use this list of settlers' names from the period as the source of names for their family members. Some of these names will appear in later activities. Seeing "their" family members in these situations can enhance students' appreciation of the history and engagement in the unit.

Mathematics

This page contains reference material for students.

Progression

Students will refer to this page during the activity *Creating Families*.

POW 6: The Haybaler Problem

Intent

As with all POWs, this offers students another opportunity to explore a problem outside of class and to communicate their thinking in writing.

Mathematics

This POW emphasizes reasoning more than any particular algebraic technique. This activity, although seemingly simple, can become quite difficult for students. As it often takes a second or even third completely renewed approach, sometimes the lesson learned is an increase in flexibility in trying methods to solve a problem.

Progression

The activity is introduced in class. Approximately one week after the POW is assigned, students engage in a class discussion of their findings.

Approximate Time

10 minutes for introduction

1 to 3 hours for activity (at home)

30 minutes for presentations

Classroom Organization

Groups, followed by whole-class presentations

Doing the Activity

The activity might best be presented without referring to the page in the student book. Tell students that you have five bales of hay and that they were weighed in pairs. Before providing the weight combinations, you might ask how many combinations there should be. Once students have identified and proven that there must be 10, provide the 10 weights given in the problem: 80, 82, 83, 84, 85, 86, 87, 88, 90, and 91.

Tell students that their task is to determine the weight of each bale of hay. If they find one solution, they are to determine whether there is more than one possible set of weights, and explain how they know.

Provide a brief amount of time for students to begin to explore. Make sure they understand that each given value represents the weight of two bales. They will need to experiment with numbers and organize their information in new ways to solve the problem.

Here are some questions you might suggest groups try to answer if they appear to be frustrated.

How much do the two lightest bales weigh? How much do the two heaviest bales weigh?

Which two bales must be weighed to get the second lightest weight?

Can you determine the weight of all the bales together?

Remind students that it is especially important to discuss their solution process in their write-ups, because the actual weights, once found, are not that interesting. They should explain how they know they have all the possible solutions, and should also look for and describe any simpler ways to approach the problem.

Select students to initiate discussion of this POW. Encourage these presenters to consider what should be in their presentations in addition to numeric answers. Presenters should consult the write-up instructions if they need ideas of what to talk about.

Discussing and Debriefing the Activity

Have the three students make their presentations. If they all found the same answers—which should be the case provided they made the same assumptions—the presentations can focus on other issues, such as the uniqueness of the solution and other methods of approaching the problem.

Ensure that audience members are listening carefully to understand the presenters' methods and thinking. Encourage them to ask clarifying questions, including questions suggested by the requirements of the activity, such as, *How did you decide this is the only solution?*

Sometimes modeling asking good questions, and explaining that you are doing so, can help students to more carefully consider the questions they might ask during a presentation; for example, *If you didn't find another solution, does that prove that there isn't another one?*

Although you might not have any students who were able to prove the uniqueness of the solution, be sure the class recognizes the distinction between not being able to find another solution and proving that there are no others.

Note that if the ten combination weights given are thought of as having been rounded to the nearest whole number—that is, are not necessarily exact sums— then there are many solutions, including some that are significantly different from the solution one gets by assuming the combination weights are exact. (Our thanks to a student for pointing this out!)

Key Questions

If the individual bales have these weights, what are the weights when they are weighed in pairs?

How much do the two lightest bales weigh? How much do the two heaviest bales weigh?

Which two bales must be weighed to get the second lightest weight?

Can you find the weight of all the bales together?

Did you assume that the weights were whole numbers? How did that affect your work?

Supplemental Activity

More Bales of Hay (extension) makes a good follow-up to presentations of *POW 6: The Haybaler Problem*. Experimentation with numbers is the essence of this activity. In particular, students are asked to consider whether such problems always have unique answers and in which cases the answers are whole numbers.

Hats for the Families

Intent

This activity builds students' skill in making sense of information presented in context, working with numeric data, and finding maximums and minimums.

Mathematics

Students look for minimal and maximal solutions that fit given numeric constraints. Because students are working with families of different sizes, they will not be able to check their work simply by comparing numeric results. Instead, they must develop a feeling for the mathematical process of obtaining their results.

Progression

This activity is meant for homework amid group work on *Creating Families*; it can also be used afterward. After working individually, students have the opportunity to verify their solutions with others.

Approximate Time

5 minutes for introduction

20 minutes for activity (at home or in class)

10 minutes for discussion

Classroom Organization

Individuals, then groups, followed by whole-class discussion

Doing the Activity

You might ask students to make an initial guess of how many people will be in all the families of the class wagon train. (This is Question 3.) Then explain that they will use the conditions—that is, the constraints—given in *Creating Families* to consider how big and how small each type of family can be. (Students do not need to have completed that activity.)

Discussing and Debriefing the Activity

Give students time to compare answers in their groups. Next, ask students to report the minimum and maximum number of members in their family units.

The summary chart posted at the conclusion of *Creating Families* can be compared with results from this activity, confirming that no family of a given kind was smaller than the minimum number possible or larger than the maximum number possible. Students can also compare their estimates on Question 3 of this activity with the actual results.

Keep in mind that there may not be clear-cut "right answers" for parts of this activity, because some of the instructions in *Creating Families* are ambiguous. Asking, **What assumptions did you make?** can draw out different solutions.

Students should recognize that once these ambiguities are resolved, Questions 1 and 2 do have exact answers. Ask for justification of these values. How can you be sure about the minimum and maximum values? The justification process makes students articulate what they figured on their own, which can be quite a challenge.

For instance, students should see that the small family must have at least three members because it has "at least one child" and "more adults than children." But they should also explain how they know that a family of exactly three people is possible, and illustrate by giving an example and explaining the relationships that the people share.

Question 3 involves estimating as to what the families for the whole class will look like. If these estimates are at all reasonable, students' work should be considered correct. For example, they might use the average of their minimum and maximum numbers from Question 2 as the average for each group, and multiply this by the number of groups in the class.

Key Questions

What are the minimum and maximum sizes you found for each type of family?

What assumptions did you make?

How can you be sure about the minimum and maximum values?

Family Constraints

Intent

In this activity, students begin using variables to represent unknown values and equations to represent and answer questions. Through context-driven, informal methods, they begin to learn that symbolic representation and manipulation have meaning.

Mathematics

This activity demonstrates some of the ways that meaning for symbolic algebra is developed in the IMP curriculum. Students start their work using variables and equations to understand numeric conditions. The focus is on the meaning of equations and the use of variables in context, and the approach to equations is intuitive. For instance, students are asked to explain what it means to represent one individual's age by C and then represent another individual's age by $C + 20$.

The activity also provides an opportunity for continuing the development of students' understanding of the distributive property.

Progression

This activity is completed individually, followed by a teacher-led discussion in which students share approaches, ideas, and unresolved questions.

Approximate Time

10 minutes for introduction

20 minutes for activity (at home or in class)

20 minutes for discussion

Classroom Organization

Individuals, followed by whole-class discussion

Doing the Activity

To introduce the activity, you may want to have students work on Question 1 and then bring the class together to share approaches and ideas. Students should understand how to use variables and constants to represent the ages of people in different generations.

Discussing and Debriefing the Activity

Bring the class together to share conclusions and to discuss unresolved problems. Here are some suggestions to focus students' thinking.

Take a solution that one person in your group came up with for Question 1a, and write an equation for it similar to the one shown in Question 1c.

Was it possible to come up with a household with two grandparents in Question 2c? Why or why not?

Calculate the average age of family members for each solution you found for Question 1. What do you notice? Can anyone explain why that happens? What about Question 2?

Ensure that the discussion makes use of the term **equation**. Also review the term algebraic **expression** and introduce the term **solution** to an equation in discussing the meaning of variables and equations.

Ask, **How can you use the equation to find the ages?** As students share various techniques, discuss the meanings they have placed on the variable C or the expressions $C + 20$ and $C + 40$. This might be elicited with the question, **What does the 20 in this expression tell us?**

Ask for replies to Question 1c: What does C mean in this equation? Record replies on the board to ensure that students understand the need to state this meaning of C as a number—in particular, the age of the child.

Bring out the connection between the equation and the situation. Emphasize that in each instance, the same number must replace the variable. Use the phrase "solving an equation" to refer to the idea of finding a number to substitute into the equation that will yield a true statement—that is, make the two sides of the equation come out numerically equal.

Ask for ideas about an equation to represent Question 3, which might have been challenging for some students. One solution might be as follows.

$$T + T + T + (T + 25) + (T + 25) + (T + 25) + (T + 25) + (T + 45) = 201$$

Select one of the equations volunteered—assuming the main difference between replies is the constant term—and determine the ages of the quadruplets. Then ask students to represent the sum of the ages of the quadruplets with an equation.

$$(T + 25) + (T + 25) + (T + 25) + (T + 25) = 128$$

To encourage an alternative expression, equivalent to the left side above by the distributive property, ask, **Can you find another way to write the sum of the ages of the quadruplets?**

Verify that all given expressions are equivalent, reminding students that any value of T entered should result in the same value. For example, when $T = 5$:

$$(T + 25) + (T + 25) + (T + 25) + (T + 25) = 120$$

$$4(T + 25) = 120$$

$$T + T + T + T + 25 + 25 + 25 + 25 = 120$$

$$4T + 100 = 120$$

Students should recognize that adding the four ages is equivalent to multiplying the age of one quadruplet by 4, and that the two processes are recorded as $(T + 25) + (T + 25) + (T + 25) + (T + 25)$ and $4(T + 25)$ and mean the same thing.

Key Questions

Can you write an equation for someone's solution to Question 1a?

Calculate the average age in the family for different solutions to Question 1. How can you use the equation to find the ages?

What does C mean in the equation in Question 1c?

Planning for the Long Journey

Intent

This activity is designed to enhance students' engagement in the unit. Groups consider what supplies might be needed during westward travels of that era and then make selections from a list of supplies to outfit their families.

Mathematics

In compiling their purchase lists, students must work with rates, which are important for the development of the concept of linear relationships. They also engage in planning, which includes such skills as organizing information and examining various arithmetic combinations.

Progression

Students work in their groups to make decisions about what their families should take on the trip. Groups work on Part I, briefly share ideas as a class, and then move on to Part II. The mathematical activity of determining costs and budgeting is followed up in the activity *Shoelaces*.

Approximate Time

25 minutes

Classroom Organization

Groups

Doing the Activity

Tell the class that they will now outfit their Overland Trail families for the journey. Point out on the map that the first leg of the westward journey took travelers from Westport, Missouri, to Fort Laramie, Wyoming. Wagon trains traveled about 20 miles per day during this part of the journey. Then ask students to consider this question: **What supply items might the travelers have needed for this leg of the journey?**

Part I: Generating Ideas

Verbally give students the instructions for Questions 1 and 2; they don't need their books for this work. Emphasize that *brainstorm* means that they are simply to offer as many suggestions as possible for consideration.

When groups are ready, bring the class together and have them report. You might compile a master list of possible supply items on chart paper and comment on any ingenious suggestions.

Part II: Making Decisions

Tell groups that they will now make decisions about certain supplies from a predetermined list. Have them examine the *Overland Trail Price List*, pointing out that this is not a complete list of the supplies that actual travelers would have needed.

You may want to suggest that groups work together on the purchases for all four of their families, but that each student has the final decision for his or her own family.

Observe groups to support organization and record keeping. In addition, listen for reasoning about how students are determining costs. This is an important opportunity to identify students who may not be fluid working with simple rates, such as 2¢/lb.

Discussing and Debriefing the Activity

You may want to have a class discussion comparing the list of brainstormed items for Part I to the *Overland Trail* Price List used in Part II. You may also want to have groups report out on the amount of gunpowder, sugar, and beans they decided to purchase for the trip.

Overland Trail Price List

Intent

Students use this information when selecting supplies in *Planning for the Long Journey*.

Mathematics

This page contains reference material for students.

Progression

Students will refer to this page during the activity *Planning for the Long Journey*. The items on this list will be referred to later in the unit.

The Search for Dry Trails

Intent

In this activity, students see that the analysis of data can help inform decision making.

Mathematics

In *The Game of Pig,* students saw that averages, particularly in the context of expected value, could be used to develop an optimal strategy for a game. Now they apply their ideas about decision making and strategy based on measures of central tendency of data, especially **median** and **mean**, with a focus on the limitations of each measure.

Progression

Students work on the activity individually and then talk about and justify their reasoning in a class discussion.

Approximate Time

5 minutes for introduction

20 minutes for activity (at home or in class)

15 minutes for discussion

Classroom Organization

Individuals, followed by whole-class discussion

Doing the Activity

Introduce this activity by stating that just as settlers did 150 years ago, students will have to make many decisions based on the information available to them. Sometimes settlers had to work with little or no information, or misinformation, or speculation, or even superstition.

In this activity, students will be asked to make decisions based on data about rainfall along various trails.

Discussing and Debriefing the Activity

Have students share their reasoning for Question 1, either in small groups or as a class. There are good reasons for choosing each of the three trails. Here are a few.

- The Smoky Hill Trail has the most consistent, and therefore the most predictable, rainfall.

- The Santa Fe Trail has the lowest mean number of rainy days and the driest individual years.

- If you throw out the first entry, the Oregon Trail has the lowest average number of rainy days, by a good margin.

Ask volunteers to comment on Addison's and Lydia's methods in Questions 2 and 3. Students might note that Addison's method tends to overemphasize Enoch's experience (42 rainy days), while Lydia's method underemphasizes this case. You might use the term *outlier* to describe a case that is substantially outside the range of the majority of values.

When appropriate, ask, **What is the mathematical name for Addison's method? For Lydia's method?** Students should be able to identify the **mean** and the **median**. (Note that neither method is necessarily the best tool for decision making in this context, even though the average was crucial in analysis of *The Game of Pig*. If students make the connection, you might confirm that **expected value** is essentially an average.)

Be sure students understand the definition of each method and how each computation is done. In particular, discuss how to find the median of a data set with an even number of items. You can use the third set of data in the table, for the Oregon Trail, to illustrate that in such a case, the median is defined as the average of the two values that "share the middle" once the items are listed in numeric order.

You might also identify mean and median as "measures of the center" for a set of data (or, more formally, "measures of central tendency"). That is, they are methods of choosing a single number to represent some type of "middle" of a data set.

Key Questions

Which trail did you choose? Why?

What is the mathematical name for Addison's method? For Lydia's method?

Setting Out with Variables

Intent

These activities lay the foundation for a meaningful understanding of symbols and their use in expressions and equations to represent relationships.

Mathematics

In *Setting Out with Variables,* students construct symbolic expressions and write summary statements that explain the meaning of these expressions in given contexts. Students also evaluate expressions using substitution and examine their equivalence using the distributive property. Finally, they use variables to write rules expressing relationships and interpret the meaning of several given rules.

If M stands for the number of men in a family, F the number of women, and C the number of children, and if B stands for the amount of water consumed by one person in a day, then the two equivalent expressions $BM + BF + BC$ and $B(M + F + C)$ can be summarized with the statement "the amount of water consumed by a family each day." That these expressions can be understood as either the sum of the water consumption of men (M), women (W), and children (C) in a day, as well as the consumption by all the people in the family ($M + F + C$) in a day, is an example of understanding the meaning behind the distributive property.

Progression

Working individually and in their groups, students track the progress of their families through a series of activities tied to places and events along the Overland Trail. In addition, they complete their work on the first POW of the unit.

Shoelaces

Laced Travelers

To Kearney by Equation

The Vermillion Crossing

Ox Expressions

Ox Expressions at Home

If I Could See This Thing

Shoelaces

Intent

Students compute the total length of shoelaces their families will need on their travels, which creates a context for bringing out methods of symbolic algebra—in particular, the use of an algebraic expression and the process of **substitution**.

Mathematics

Students extend their work with variables and algebraic expressions in *Family Constraints* to examine how to use variables to represent complex arithmetic algorithms. Students are introduced to the formal process of substitution into and evaluation of algebraic expressions, emphasizing the two steps in the process.

Progression

Students work in small groups on their own Overland Trail families, while listening to classmate's methods for figuring the total length of shoelaces needed. Students share verbal and written statements about procedures for calculating shoelace length. The teacher leads a discussion to convert the generalization to an algebraic expression and test it by substitution.

Approximate Time

25 minutes

Classroom Organization

Individuals, in collaboration with group members, followed by whole-class discussion

Materials

Information about each group's four families

Doing the Activity

Transition students from their work on *Planning for the Long Journey* by asking whether anyone thought about some of the more mundane supplies the travelers would need. In this activity, students figure the total length of shoelaces their families may need along the trail.

This is a good opportunity to assess how group interactions are progressing. Each group member should arrive at the same solution to Questions 1–3, but a different solution to Question 4. (FYI, a woman needs 224 inches of shoelace, a man 256 inches, and a child 144 inches.)

You might interrupt groups to encourage students to check in with each other and to listen to the reasoning behind each student's solution to Question 4. Ask, **What similarities do you see among your methods for Question 4?** Conversation about their methods will likely help students answer Question 5 prior to the class discussion.

Encourage groups that finish early to figure the amount of shoelace needed for another person's family, or the whole group's families, or the entire class's families. They might also write a general rule—in written form, as a set of steps, or algebraically— to calculate the amount of shoelace needed for a family of any size.

Discussing and Debriefing the Activity

Have two or three students present their work on Question 4. Since each student is discussing a different Overland Trail family, their numeric answers will differ.

To transition to Question 5, ask a presenter, **Describe in words how you found your answer to Question 4.**

Question 5 is really the heart of the activity, as it leads to the generalization into variables. You might have a couple of additional volunteers show their descriptions. It's useful for students to see more than one description, even if the descriptions reflect the same arithmetic process. You will probably get descriptions like the following.

Multiply the number of women by 224, the number of men by 256, and the number of children by 144, and then add these products.

Ask students whether they see an opportunity to abbreviate these verbal descriptions by using variables. **Does anyone have a suggestion for abbreviating these verbal descriptions by using variables?** If necessary, remind students of how they used variables in *Family Constraints*. Insist that students give clear definitions for the letters they introduce, such as, "W represents the number of women in the family" instead of "W equals women."

The class should be able to come up with an expression like $224W + 256M + 144C$ for the amount of shoelace needed for a family with W women, M men, and C children.

Post this expression and the equivalent verbal description, which will be a useful reminder for students, particularly during the activity *Ox Expressions*.

Remind students that such a combination of numbers and letters is called an *algebraic expression*. Review the term **coefficient** and remind students of the convention to omit the multiplication sign in expressions like $224W$. Take a moment to ensure that students understand that these abbreviations are simply notational conventions; that is, agreements among mathematicians to write things a certain way. There is nothing inherently wrong about placing a multiplication sign between a number and a variable or about placing a variable in front of the coefficient—it is simply not the common way of communicating rules in mathematics. You might mention order of operations as another example of a mathematical convention.

Provide some examples to illustrate how a general algebraic expression is used to evaluate a particular case. For example, ask students to use the expression $224W + 256M + 144C$ to find the amount of shoelace needed for a family with four women, six men, and five children.

It is often helpful to identify two stages in the process.

- *Replacing a variable by the desired numeric value.* In this example, this means writing out the expression $224 \cdot 4 + 256 \cdot 6 + 144 \cdot 5$. Be sure students realize that they need to restore the multiplication signs that were omitted when letters were used.

- *Carrying out the indicated arithmetic using order of operations.* This means, for example, finding that the numeric expression $224 \cdot 4 + 256 \cdot 6 + 144 \cdot 5$ is equal to 3152.

The two stages are sometimes called *substitution* and *evaluation,* although either term can refer to the overall process.

Key Questions

What similarities do you see between each of the methods for answering Question 4?

How much shoelace did your Overland Trail family need?

Describe in words how you found your answer to Question 4.

Does anyone have a suggestion for abbreviating these verbal descriptions by using variables?

How can you use the expression $224W + 256M + 144C$ to find the amount of shoelace needed for a family with four women, six men, and five children?

Supplemental Activities

Pick Any Answer (reinforcement) provides a simple context for which students can create an algebraic expression. It might be an appropriate activity to use in connection with students' work here with variables.

Substitute, Substitute (reinforcement) provides examples through which students can strengthen their understanding of the basic ideas of substitution. The activity uses a variety of phrases for referring to the process. One question uses substitution to reinforce the idea of combining terms.

Laced Travelers

Intent

In this activity, similar to their work in *Shoelaces*, students put arithmetic processes into words.

Mathematics

Students work through and describe the arithmetic used to solve several problems, setting the stage for writing algebraic expressions to describe situations.

Progression

After working on their own, students focus on the verbal descriptions of their computations. Class discussion is oriented once again, as in *Shoelaces*, to translating verbal descriptions into algebraic expressions. This activity leads to more practice with substitution in *To Kearny by Equation* and sets the stage for students to focus on the meaning of algebraic expressions in *Ox Expressions.*

Approximate Time

10 minutes for introduction

20 minutes for activity (at home or in class)

15 minutes for discussion

Classroom Organization

Individuals, followed by whole-class discussion

Doing the Activity

Introduce the activity, reminding students to use the new data that are given, rather than the information in *Shoelaces*.

Discussing and Debriefing the Activity

You might begin by having students share the questions they made up for Question 2 and try to answer each other's questions. Meanwhile, ask two or three students to put their verbal descriptions from Question 1 on transparencies.

The main focus of the discussion should be on turning students' verbal descriptions into algebraic expressions. For example, suppose one student's description for Question 1 begins, "You take the 150 wagon trains times the 25 families per train times the length of shoelace per family. The amount per family is twice the amount per man plus the amount per woman plus three times the amount per child." Ask students to rewrite this sentence without reference to specific numbers. For example, the first sentence would come out something like, "Multiply the number of wagon trains by the number of families per train by the length of shoelace per family."

To write this as an algebraic expression, students need to choose a letter to represent each quantity, such as *T* for the number of wagon trains, *F* for the number of families per wagon train, and *S* for the total amount of shoelace needed per family. The total amount of shoelace needed for an entire year would then be represented by an expression like *TFS*.

Emphasize the difference between an object and the number of such objects and the importance of being precise about what a letter represents. For instance, in the example above, *F* does not represent a family, or the number of people in a family, but *the number of families in a wagon train*.

As time allows, build on the discussion by having students share some of the questions they wrote for Question 2 and their verbal descriptions from Question 3, and having the class develop algebraic expressions to go with these questions.

Key Questions

Describe in words how you found the answer.

How can you rewrite this sentence without reference to specific numbers?

How can you use variables to rewrite this sentence as an algebraic expression?

Supplemental Activity

From Numbers to Symbols and Back Again (extension) uses formulas from two settings from earlier units (*The Game of Pig* and *Patterns*) as the context for work with substitution. Students will have to guess and test solutions to Questions 1c and 2b, as the equations are quadratic. Furthermore, the solution to Question 1c is not integral (or even rational), so students will have to determine a reasonable approximation.

To Kearney by Equation

Intent

This activity creates an opportunity for further development of symbolic algebra.

Mathematics

Students begin with a formula for profit and are asked to explain why it makes sense. They then adjust the formula to represent half the total profit, a process that can lead to two equivalent expressions and create a context for bringing forth the idea of equivalence and the distributive property. Students conclude the activity with substitution and evaluation practice.

Progression

Students work individually and in their groups on the task, coming together as a class to review the ideas of equivalence and the distributive property.

Approximate Time

25 minutes

Classroom Organization

Individuals, then small groups, followed by a brief whole-class discussion

Materials

Information about each group's four families

Doing the Activity

Locate Kearny, Nebraska, on the map so students can connect this activity to their journey westward from Westport. Also locate the Kansas River, in Topeka, Kansas, along the route to Kearny. If a student (or you) has ever visited Kansas or Nebraska, talk about their geography. Remind students that 150 years ago, there were no cars, highways, or bridges. Students might speculate about how supplies, wagons, and animals were transported across rivers.

Introduce the activity, letting students know that Joseph and Louis Papan were real individuals.

As groups work on Questions 3 and 4, watch for common sources of confusion that may warrant a class discussion. Disagreements about the numeric answers can arise from at least two sources:

- Incorrect values substituted for the individual variables—for example, students may have forgotten to account for some families having more than one wagon

- Arithmetic mistakes, possibly due to misuse of order-of-operations rules

Discussing and Debriefing the Activity

In the discussion, take advantage of the opportunity to review, once again, the distributive property.

For Question 2, students may have identified two ways to split the profit.

- Split the $2 per wagon to get $1 for each brother, and, separately, split the 30¢ per hour pay so they each pay 15¢ per hour. This gives the expression $W - 0.15H$ for each brother's profit.

- Split the overall profit between the brothers so each gets $\dfrac{2W - 0.3H}{2}$.

If both methods arise (you may want to suggest the one that does not), use the opportunity to speak about the distributive property. The equivalent results illustrate that $W - 0.15H$ and $\dfrac{2W - 0.3H}{2}$ must represent the same amount and so must be equivalent expressions.

The Vermillion Crossing

Intent

This activity will offer information about how well students can interpret symbolic expressions and perform substitution.

Mathematics

Students are asked to explain how a new formula for profit—a linear function of the number of wagons, men, women, and children—makes sense. They conclude the activity with substitution and evaluation practice.

Progression

Students will work on this activity individually, followed by a class discussion of various approaches.

Approximate Time

10 minutes for introduction

20 minutes for activity (at home or in class)

10 minutes for discussion

Classroom Organization

Individuals, then groups, followed by whole-class discussion

Materials

Information about each group's four families

Doing the Activity

Tell students that the ferryboat operator in this activity, Louis Vieux, was a real individual. Pointing out the Vermillion River on a map can lend more meaning to the context. As this river is a bit tricky to locate, doing so could be a nice extension for some students.

Discussing and Debriefing the Activity

Have students share answers in their groups. The class might then discuss how the various groups' total costs vary.

Ox Expressions

Intent

Students work with a given set of variables to generate meaningful algebraic expressions. The approach emphasizes the context in which the expressions emerge, giving students a meaningful connection to the symbols, equivalence, and symbol manipulation.

Mathematics

Students have created algebraic expressions to describe calculation procedures. Now they will move in the other direction, interpreting algebraic expressions in terms of the concepts that the variables represent. Students explain and evaluate algebraic expressions, generate meaningful algebraic expressions, and interpret algebraic expressions using **summary phrases**. The equivalent meaning of the expressions developed with a given summary phrase will help students to recognize that the expressions are equivalent, offering another opportunity to identify the distributive property at work.

Progression

Students work in groups to create meaningful algebraic expressions and describe them with summary phrases. The activity concludes with sharing findings and discussing the experience of writing algebraic expressions and summary phrases.

Approximate Time

30 minutes

Classroom Organization

Groups, followed by whole-class discussion

Doing the Activity

This activity asks students to write as many different meaningful algebraic expressions as possible for a given set of variables. The key idea is to give contextual meaning to algebraic expressions, rather than seeing them simply as descriptions of a set of arithmetic operations.

Have students read the activity. Discuss the idea of a summary phrase for describing an algebraic expression, emphasizing that summary phrases should be as concise as possible. For instance, using the example given in the activity, the summary phrase for the expression FC is "the number of children traveling in a train" rather than "the number of families traveling in a train times the number of children in a family."

Also emphasize, as mentioned in the activity, that not every algebraic expression has meaning in the given context. You might illustrate with some additional examples or ask students to create one or two of their own.

Discussing and Debriefing the Activity

If the students in a group seem overwhelmed after working a bit, ask them to begin by considering just three symbols: *M* (the number of men in a family), *W* (the number of women in a family), and *C* (the number of children in a family). Then ask what the expression *M* + *W* + *C* represents in terms of the problem (the total number of people in a family). Similarly, ask about the expression *M* + *W* (the number of adults).

You can then introduce other variables, such as *F*, *B*, and *D*. Ask, **What questions can you express using these variables?** The group can brainstorm to come up with questions that use those variables, such as "How much water do the women in the wagon train drink altogether on the trip?"

It is unlikely that the whole class will need such support to move forward. If several groups are stuck, though, another strategy is to call up a representative from each group to create an algebraic expression for the summary phrase, "The amount of water the women in the wagon train will drink altogether on the trip."

You might notice students using the numeric values in their work. Some students are quite productive this way, using the numbers to confirm that their expressions make sense. Others might be tricked into thinking that simply because the numbers "work" the expressions make sense.

Bring students together to create a class list of meaningful "ox expressions." You might ask each group to give one of its expressions together with a summary phrase. Other groups can then challenge the meaningfulness of the phrase or propose an improvement.

One error students often make is to neglect to say "number of" in defining variables. For example, they might say "*F* is the families in a wagon train" rather than "*F* is the number of families in a wagon train." Pay attention to this.

After generating a long list of expressions, you may want to discuss these questions.

Which letters appear most frequently in the expressions? Why is that?

What determines whether an expression has meaning or not?

Do you think you found all the meaningful expressions? Why or why not?

Look for opportunities to point out the use of the distributive property, possibly posting any examples.

Key Questions

What is a summary phrase for the expression FC?

What does $M + W + C$ represent in terms of the problem?

What questions can you express using these variables?

What is one of your expressions, and what is the summary phrase that goes with it?

Which letters appear most frequently in the expressions?

What makes an expression meaningful?

Do you think you found all the meaningful expressions?

Ox Expressions at Home

Intent

This activity will help you assess how well students understand meaningful algebraic expressions.

Mathematics

As in *Ox Expressions*, students approach the use of variables from another perspective, generating meaningful algebraic expressions from a given set of variables. They see that, when used in context, combinations of variables can represent more than a sequence of numeric steps. Students also strengthen their skill with substitution into algebraic expressions. Finally, the activity offers another context for working with the distributive property.

Progression

Students complete the activity on their own. After this individual work, while students share their expressions and summary phrases, the teacher can gauge their understanding to determine what to debrief and what to emphasize. The activity concludes with identification of examples of the distributive property.

Approximate Time

10 minutes for introduction
20 minutes for activity (at home)
15 minutes for discussion

Classroom Organization

Individuals, followed by whole-class discussion

Materials

Sentence strips (optional)

Doing the Activity

This activity is similar to *Ox Expressions*, but more structured. One difference is that the numeric values have been stripped from the list of variables. You might watch how students adapt their work to this change and inquire whether this makes the activity easier or more challenging.

Tell students that in this activity, they will work on problems similar to those in *Ox Expressions*, using algebraic expressions and summary phrases.

Discussing and Debriefing the Activity

You might begin by having students share in their groups. While you are listening, watch for students who write statements of arithmetic procedures rather than summary statements. For example, in Question 5, students should recognize that *FM* gives the number of men in a wagon train and should not simply describe the expression as the number of families in a wagon train times the number of men per family.

In Question 7, students should recognize that the expression *WL* has no meaning in this context, even though one can multiply the numbers that might be associated with the variables *W* and *L*.

The extent of the discussion will depend on what happened in the earlier discussion of *Ox Expressions*. If students had difficulty with that, continue to emphasize the issue of meaningful expressions and the creation of summary phrases. The suggested supplemental activities can help students who need more experience with relating variables to their meanings and creating good summary phrases.

This activity also lends itself to using sentence strips, or another public display, of student replies. Responses to Question 8 can be posted as challenge tasks—posting only the ox expression or the summary phrase, of course.

Use the discussion of this activity to reinforce students' understanding of the distributive property. For example, focus on Question 2 to bring out that the solution can be written as either $B(W + M + C)$ or $BW + BM + BC$, and identify this as another example of the distributive property. If either of these expressions did not arise in students' work, challenge the class to seek another expression for the summary phrase "the water consumed in a day by a family." Ask, **How could you get the total amount of water by finding the separate totals for the women, for the men, and for the children?**

Ask whether students can explain why the expressions $B(W + M + C)$ and $BW + BM + BC$ are both correct. Use this opportunity to review the phrase *equivalent expressions* and to bring out that the equivalence of the expressions is an example of the distributive property. You might also suggest that students assign specific numbers to the variables and use substitution to help confirm the equivalence of the two expressions. Having different groups use different values will make the confirmation even more powerful.

Key Question

How could you get the total amount of water by finding the separate totals for the women, for the men, and for the children?

Supplemental Activities

Classroom Expressions (reinforcement) asks students to create summary phrases and algebraic expressions using a set of variables that relate to a classroom setting. The activity also introduces the notation of subscripts.

Variables of Your Own (reinforcement) asks students to make up a set of variables, and to write algebraic expressions and summary phrases, in a context of their own choosing.

If I Could See This Thing

Intent

This activity extends students' work with writing expressions for mathematical situations.

Mathematics

In this activity, students are asked to create equations to represent the relationship between future and present populations. They are encouraged to use numeric examples to clarify their use of variables. The equivalent expressions that emerge offer another opportunity to review the distributive property.

Progression

Students work individually and in their groups before coming together as a class to share their thinking about converting from written problems to algebraic equations.

Approximate Time

10 minutes for introduction

20 minutes for activity (at home or in class)

15 minutes for discussion

Classroom Organization

Individuals, then small groups, followed by whole-class discussion

Materials

Information about each group's four families

Doing the Activity

Read the passage from *Life of George Bent* aloud with students before they start their work. On the map, point out the part of the trail to which the activity refers—from Fort Kearny, Nebraska, to Fort Laramie, Wyoming, along the North Platte River.

If the activity is assigned for homework, it may be worth having students spend time in their groups to do at least one numeric example for Question 1 to ensure that everyone realizes that the process will involve multiplying the population by the decimal equivalent of the percentage of loss. As necessary, remind students of the difference between an expression and an equation.

This activity might be difficult for many students, primarily because they may be unfamiliar with what it means for a population to decrease by a given percentage.

The suggestion in the student book to begin with a particular number to represent the starting population is useful, but students may still need encouragement to complete the necessary mathematics. Once a student has selected a population for 1492, you might ask, **How much is 90% of that population?**

Although the computations in the activity are simple (involving multiplication and subtraction), the process of using variables and equations to represent information about real-world situations can be challenging. As this is a fundamental idea in mathematics, allow ample time for questions and discussion. Students may also need help identifying the variables in the situations.

Students will need to know the total number of adults and total number of children on the class wagon train. They should have this information from their work on *Overland Trail Families*.

Discussing and Debriefing the Activity

Have students share ideas in their groups as you assess what they have been able to accomplish with the task.

Ask the students presenting Question 1 to describe exactly what they did, especially the process by which they arrived at an equation. Most students will use one variable (for example, B) for the population at the *beginning* of the time period and another (for example, E) for the population at the *end* of the time period and write an equation such as $E = 0.1B$ or $E = B - .09B$ to express the relationship between them. (Estimates of the Native American population in 1492 vary widely, from 800,000 to 30,000,000. The 1900 census reported the Native American population as 237,000.)

Next, have students share ideas about creating a rule for the revised wagon train size in Question 2. This is more difficult than Question 1 because it involves two input variables.

Students may have had trouble finding a rule for their tables in Question 2c. If so, have them describe what they did to get their numeric results in Questions 2a and 2b, which can help them to see an arithmetic pattern and produce an equation like $N = 0935A + 0.9C$ to describe the situation. Be aware that some students may not yet be completely comfortable with such algebraic symbolism.

Someone may point out that it doesn't make sense for the output from this table to be anything other than a whole number. You can let the class come up with a way to resolve this dilemma, such as rounding off to the next highest integer.

Key Questions

What did you do to come up with an equation for Question 1?

How did you get the numeric results in Questions 2a and 2b?

Supplemental Activities

Painting the General Cube (extension) is a challenging activity that asks students to create equations to describe a geometric situation.

Integers Only (extension) introduces the *greatest integer function*. Interested students might try to use this function to devise a formula for Question 2c of *If I Could See This Thing*.

The Graph Tells a Story

Intent

The activities in *The Graph Tells a Story* are devoted to the development of graphing as a mathematical tool and link this tool to equations, tables, and situations.

Mathematics

A function may be represented four main ways: as an In-Out table, as a symbolic rule, as a graph, or verbally as a situation. In *The Graph Tells a Story*, the focus is on graphical representations of functions. Sketches of graphs, with unlabeled axes, can provide a sense of the overall behavior of a relationship, while carefully drawn graphs with labeled axes can summarize a great deal of information about the specifics of a relationship. A graph's axes define a coordinate system; points in the plane have coordinates that correspond to the rows of an In-Out table. Patterns in an In-Out table correspond to the shape of the graph and the form of the symbolic rule. Pairs of numbers that make the rule true correspond to points on the graph and rows in the table. The groundwork for this fluent movement among a function's representations is the focus of these activities.

Progression

The activities begin with students interpreting graphs in a qualitative way, looking at their overall shape rather than specific points. Students then interpret graphs quantitatively, creating scales and reading coordinates of specific points. Finally, they link graphs with tables, symbols, and situations. In addition, students begin work on the unit's second POW.

Wagon Train Sketches and Situations

Graph Sketches

In Need of Numbers

The Issues Involved

Out Numbered

From Rules to Graphs

POW 7: Around the Horn

You're the Storyteller: From Rules to Situations

Wagon Train Sketches and Situations

Intent

This activity sets the groundwork for students to learn the techniques of graphing and the connections among graphs and other representations of functions.

Mathematics

Students begin to look at graphical representations of the relationship between two quantities, developing meaningful connections between a graph and the situation it represents. They interpret and create graph sketches for given situations. They recognize the distinction between discrete and continuous graphs and identify dependent and independent variables. This exploration sets the stage for the need to meaningfully scale graphs, followed by plotting points and graphing equations.

Students will be interpreting graphs in a qualitative way, focusing on rise and fall. They will also create sketches of graphs to reflect information describing various relationships. (The activity uses the term *graph sketches* to describe these more qualitative graphs.) In later activities, they will add scales, plot points, and draw graphs of equations on coordinate axes.

Progression

After the teacher introduces the activity, students work in groups, coming together as a class to share observations about the graphs. The teacher will use these observations to introduce such vocabulary as *constant rate,* and **linear**, **discrete**, and **continuous graphs**. The teacher also introduces **independent** and **dependent variables** and the graphing convention for placement of these variables on the *x*- and *y*-axes, respectively.

Approximate Time

30 minutes

Classroom Organization

Groups, followed by whole-class discussion

Materials

Wagon Train Sketches and Situations blackline master (transparency)

Doing the Activity

This activity assumes that students are generally familiar with this type of graph and the use of vertical and horizontal axes. You may want to mention that these graphs are quite different from the frequency bar graphs they worked with in *The*

Game of Pig. During the activity, use the terms **vertical** and **horizontal axes** as students refer to the two perpendicular lines in their graph sketches.

To introduce the activity, either read through the brief introduction with students or do the example at the front of the room and introduce the task verbally.

You may want to acknowledge that the axes of the graph sketches do not have scales, but that students should still think of them as representing numeric information, with the numbers increasing as one moves up or to the right.

Ask students, **What does the "shoes or boots versus people" graph tell you?**

There are at least two observations they are likely to make.

- The more people there are, the more shoes or boots are needed.

- The relationship is linear. That is, the dots lie in a straight line.

For now, the term *linear* should be understood simply in its geometric sense: the graph follows a straight line.

Have students begin work in their groups on the activity, reminding them to keep a written record of their ideas. As you circulate, ask groups for explanations of the phenomena described in the graphs that students may have overlooked or what might be happening to cause a graph to "behave" in a certain way.

Discussing and Debriefing the Activity

You need not discuss all of the questions, although it may be fruitful to have students give short presentations for some of them. Following are some ideas to look for in a discussion or as you circulate among groups.

Part I: From Sketch to Situation

Question 1: Students should be able to deduce from the linearity of this graph sketch that coffee consumption is the same each day. They may express this by saying that the graph "goes down" the same amount each day. Informally use the term *constant rate* in the context of this problem. The concept of rate is fundamental to the work of this unit, and the word will be used in many settings.

Question 2: Students may have various theories about what is happening in this graph sketch. One reasonable theory about the horizontal portions of the graph is that the wagon train group was traveling along a river and got water from the river rather than from the water barrels. So point *A* would represent arrival at the river, point *B* would represent people filling up their barrels, and point *C* would represent a time when they were traveling along another river.

Questions 3: Students should be able to articulate that the wagon train is moving fastest when the graph is "steepest." You can ask what might cause changes in the wagon train's speed (terrain, weather, and so forth). You can also ask what the graph sketch would look like if the wagon train had covered the same distance at a constant speed.

Question 4: This graph consists of individual points because the number of wagons must be a whole number. You might also point out that the points lie on a straight line and ask why that makes sense. Introduce the term **discrete** for a graph that consists of individual points.

You can also bring out the contrast between the graph in Question 4 and the graph in Question 3, which is unbroken because the time and distance concepts make sense for any nonwhole-number values. Introduce the term **continuous** for an unbroken graph. Reinforce the vocabulary by asking which of the other graphs in the activity are discrete and which are continuous.

Part II: From Situation to Sketch

Ask a couple of volunteers to present their graph sketches for each situation. Encourage students to make observations about the mathematics of their work.

Questions 5: Students might notice that this graph is discrete because the number of wagons is a whole number. The points on the graph should lie on a straight line because the wagons are "of a fixed size and type."

Question 6: No specific information is given about the rate at which the buffalo population decreases as the number of settlers increases, so any decreasing graph is reasonable. Because the numbers are likely to be large, this graph might be treated as continuous, even though, technically, it would be discrete.

Question 7: The key idea here is that the graph should start positive, decrease to zero, and then increase. If the rider is assumed to be going at a constant speed, the graph should have the V shape of the **absolute value** function. As an extension, you might ask students to compare their graphs for this question with a graph of how far the rider has traveled as a function of time, and use that comparison to talk about the idea of absolute value.

One issue that may arise in the discussion of this activity is which axis to use for which variable—if not, raise it yourself. Explain that in many situations, one of the variables is *dependent* on the other. For example, in the graph introducing the activity, the number of pairs of shoes or boots needed *depends on* how many people there are in a family. (It may depend on other things as well, but that's another issue.) Introduce the terms **independent variable** and **dependent variable**, and tell students that these concepts are essentially synonymous with the input and output of an In-Out table. Also mention that the convention in mathematics is to put the independent variable along the horizontal axis and the dependent variable along the vertical axis.

Key Question

What does the "shoes or boots versus people" graph tell you?

Supplemental Activity

Spilling the Beans (reinforcement or extension) is an activity with a western setting. Although the mathematics is more about reasoning—and specifically about proportional reasoning—the context makes this an appropriate place to use this activity.

Graph Sketches

Intent

This activity will give you a sense of how well students can interpret ideas about the relationship between two variables as communicated in a graph. The activity also sets the stage for assigning numbers to the axes of a graph.

Mathematics

To interpret the "story" that a graph tells and to create a graph to represent a story, students must focus on how a relationship between two quantities can be expressed in this visual form. Along the way, they are encouraged to continue to employ the language of graphing—in particular, independent variable and dependent variable.

Progression

Following their explanation of *Wagon Train Sketches and Situations,* students work individually to interpret graphs and then to create their own graphs. In a class discussion, they share ideas and interpret each other's work.

Approximate Time

5 minutes for introduction

20 minutes for activity (at home or in class)

15 minutes for discussion

Classroom Organization

Individuals, then whole-class discussion

Materials

Graph Sketches blackline master (transparency)

Index cards (optional)

Doing the Activity

When you introduce the activity, you may want to suggest that students put their work for Part II on index cards to facilitate an exchange of problems in the discussion.

Discussing and Debriefing the Activity

Allow time for students to share their ideas in their groups about the graphs in Part I.

You may want to introduce the term *step function* for the situation in Question 3. This term describes a function whose graph jumps from one horizontal segment to the next.

You might use Questions 2 and 4 to distinguish between linear and nonlinear situations. As needed, help students to articulate that in Question 2, the graph indicates that the student's "rate of work" increases as the POW deadline gets closer, whereas in Question 4, the amount of money grows at the same rate with each ticket sold.

You could also use these examples to illustrate discrete versus continuous variables and their resulting graphs. The independent variables in Questions 2 and 4 take on only whole-number values, whereas the dependent variables can be interpreted in different ways, depending on the degree of precision. In contrast, Question 3 has a continuous independent variable and a discrete dependent variable, while Question 1 is continuous in both variables. This idea will arise again in Question 4 of *The Issues Involved*.

In Part II, students may have imaginative descriptions to accompany their sketches. You might have each group pass their papers (or index cards) to the next group, with the descriptions face up. Each group should attempt to make sketches that illustrate the descriptions and then compare their sketches with those provided by the creators of the problems. If disagreement arises about whose answer is "correct," emphasize that there can often be more than one correct graphical interpretation of a verbal description.

Supplemental Activity

More Graph Sketches (reinforcement) provides a variety of contexts for which students can create additional graph sketches.

In Need of Numbers

Intent

In this activity, students are asked to add reasonable numeric scales to the axes of graph sketches from previous activities. The key idea is that each axis is to be treated as a number line, or part of one.

Mathematics

This activity continues the development of graphing strategies by introducing the process of quantifying graphs using scales. Students begin to understand that each axis is a number line. They read information from graphs and use this information to add scales to graph sketches. Students must justify the reasonableness of their scales. In the process, they employ basic ideas and terminology about coordinate graphing, in particular the term *coordinate*. Scaling the axes of a graph and reading quantitative information from graphs are conceptual sticking points for many algebra students, so the groundwork laid in this activity is particularly important for the mathematical development of the unit.

Progression

After students have shared ideas from *Graph Sketches*, the teacher introduces the process of scaling the axes on a graph. Students then do the activity in groups and begin to wrestle with the difficulties associated with scaling axes. *The Issues Involved* is an immediate follow-up activity in which issues about scaling are raised and more conventions for graphing are defined.

Approximate Time

25 minutes

Classroom Organization

Whole-class introduction, followed by small groups

Materials

In Need of Numbers blackline master (transparency)

Doing the Activity

You might use the graph from Question 1 of *Wagon Train Sketches and Situations* to demonstrate the level of thought and detail you expect in this activity. The initial, unscaled graph looks like this.

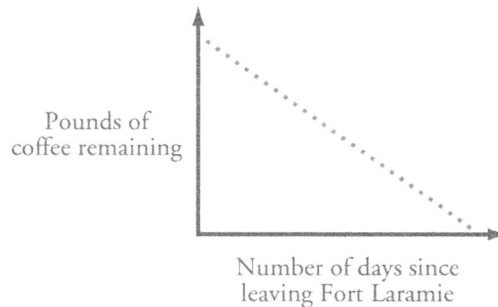

To choose appropriate scales, students will need to make estimates. They might start by estimating how much coffee the family might have begun with and how long it might have taken for them to use it up. For example, if the family started with 25 pounds of coffee and consumed half a pound of coffee per day (perhaps for a large Overland Trail family), the family will have run out of coffee after 50 days. For those values, the graph with scaled axes might look like this.

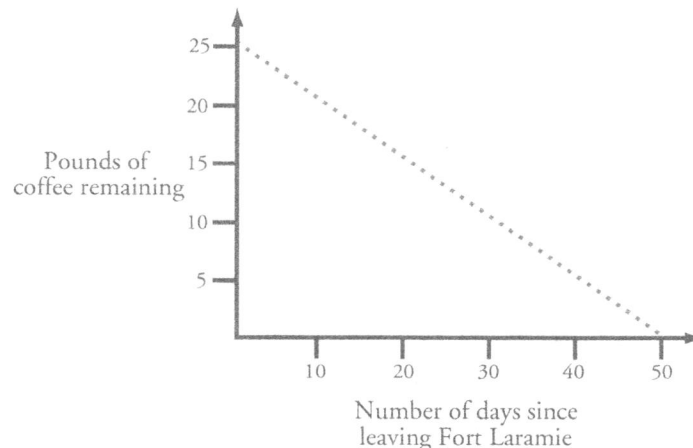

You can ask whether it's permissible for "10 days" and "10 pounds" to be represented by different lengths on the two axes. In general, if the units on the two scales are not measuring the same type of item, there is no reason for the lengths to match.

Once the class has decided how to scale the axes, ask some questions that can be answered from the graph, such as, **How much coffee is left after 10 days? After 20 days?**

Have students both articulate and demonstrate how they are finding these answers. For instance, to determine how much coffee is left after 20 days, they would locate 20 on the horizontal axis, go up (vertically) from there to the graphed line, and then go over (horizontally) to the vertical axis and read off the value (in this case, 15).

Identify these two numbers, 20 and 15, as the *coordinates* of the associated point on the graph. Also bring out the units associated with these numbers: this point on the graph represents that after 20 *days* since leaving Fort Laramie, the family has 15 *pounds of coffee* remaining. You can introduce ordered pair notation (20, 15) now or delay this until the activity *The Issues Involved*.

Discussing and Debriefing the Activity

Circulate as groups work, using the opportunity to encourage the use of successful group strategies, such as collaboration and checking in with one another.

This activity will not need a whole-class debriefing. The next activity, *The Issues Involved,* raises general issues about creating scales that may lead the class to look back at the graph sketches from this activity.

Key Questions

What would be an appropriate scale for this axis?

Is it okay if "10 days" and "10 pounds" are represented by different lengths on the axes?

How much coffee is left after 10 days? After 20 days? How do you find these values from the graph?

The Issues Involved

Intent

In this activity, students examine issues related to the scaling of graphs.

Mathematics

This activity raises some general issues about the scaling of graphs, in particular that scales should be consistent—evenly spaced—and that axes are usually assumed to begin at zero unless otherwise marked. The discussion of the activity offers opportunities to emphasize important vocabulary—such as **continuous graph**, **discrete graph**, *first coordinate, second coordinate, coordinate axes,* and **ordered pair**—and to relate graphs back to In-Out tables.

Progression

After students work individually, the teacher leads a discussion to clarify ideas, using student observations to communicate the conventions associated with scaling graphs.

Approximate Time

5 minutes for introduction

20 minutes for activity (at home or in class)

20 minutes for discussion

Classroom Organization

Individuals, then small groups, followed by whole-class discussion

Materials

The Issues Involved blackline master (transparency)

Doing the Activity

Tell students that in this activity they will consider the issues involved in making good decisions about scaling the axes of a graph.

Discussing and Debriefing the Activity

Have students share their questions and problems from Question 1 in their groups and try to answer each other's questions. At the same time, assign each group one of Questions 2 through 4 to present to the class.

After a short time, bring the class together and ask each group to state one of the questions or difficulties relating to the scaling of axes that they encountered. If the group resolved the question, ask that they share the answer or solution.

In the discussion of Question 2a, explain that there is no absolute rule for whether the vertical or horizontal axis should begin at zero; it is really a judgment call. In some cases starting a scale other than at zero gives a wrong impression of the data, while in others it does not. It is also true that for some data sets, starting the scale at zero might give the wrong impression. You may want to mention the ethical issues of using misleading scales, which are discussed in the next unit, *The Pit and the Pendulum,* in the activity *A Picture Is Worth a Thousand Words*.

In Question 2b, elicit the idea that there needs to be enough information shown to give a sense of scale, but beyond that, it is a judgment call as to what would make a graph more readable and what would clutter it up. Make sure students realize that they do not have to include every integer value along a particular axis.

In Question 3, students should be able to articulate that the conclusion that boys grow at a constant rate through age five is faulty; the uneven labeling along the vertical axis distorts the graph. In simple Cartesian graphs like those in this unit, the scales should be marked so that equal distances on a given axis correspond to equal amounts, just as on a simple number line. This is a different issue from whether the scale on the vertical axis should be the same as the scale on the horizontal axis.

You might ask explicitly how much an average boy grows in his first year, his second year, and so on (based on this graph), both to focus on students' graph-reading skills and to clarify that the amount decreases from year to year.

A correct graph for Question 3 might look like this. This graph makes it clear that a child's growth generally slows down as the child gets older.

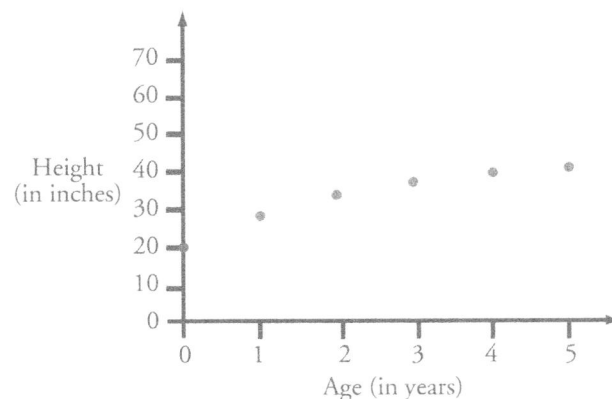

In Question 4, students should become aware that the issue of whether a graph is discrete or continuous usually depends on the context of the problem. Use this occasion to review the two terms, if necessary, and have students share their reasoning about whether this graph should be continuous or discrete. They may be able to draw larger generalizations, such as, "If the independent variable represents something that is counted with whole numbers, the graph will consist of individual points."

Caution students that sometimes graphs that should be discrete are presented as continuous. This may be because the scale used makes it impossible to draw so many dots, because the graph is more readable without dots, or just because it simplifies the problem.

During the course of the discussion, important mathematical vocabulary can arise naturally as students need or use a term. As needed, remind students that the two numbers associated with a point on a graph are called its coordinates, with the number from the horizontal axis called the *first coordinate* and the number from the vertical axis called the *second coordinate*. The vertical and horizontal axes are sometimes referred to as coordinate axes.

Explain that we often represent a point on a graph by giving its two coordinates. For instance, in the "height versus age" graph, it appears that at 1 year old, an average boy is 30 inches in height. There is a point on the graph that corresponds to this information, and the conventional way to represent that point is (1, 30). Tell students that this notation is called an **ordered pair**. Ask whether they can think of a reason why the word *ordered* is used in this expression.

Also bring out that looking at a graph is like looking at an In-Out table: each point on a graph represents a pair of numbers, as does each row of an In-Out table. Remind students that we generally associate the horizontal axis with the *In* and the vertical axis with the *Out*.

Key Questions

What problems or questions did you have scaling the axes? How did you resolve them?

According to this graph, how much does an average boy grow in his first year? In his second year?

Out Numbered

Intent

This activity will help students draw connections between what they know about reading information from graphs and what they know about In-Out tables and rules. In addition, several standard terms and conventions are introduced. The activity prepares students to reflect on how to move from rules to graphs, the focus of the next activity.

Mathematics

This activity will further the connections students are building between situations, graphs, tables, and rules. Students read data from scaled graphs of linear situations and find rules for the data, pausing to consider that the straightness of a graph corresponds to a constant rate of change. They also review the four-quadrant coordinate system, using the terms rectangular (or Cartesian) coordinate system and **quadrant**.

Progression

Students work on the activities in groups, sharing ideas about how to respond to the questions. Sometime after groups have completed Question 2, the teacher brings the class together to discuss observations, highlighting the mathematical connections students observe between representations and introducing new vocabulary.

Approximate Time

40 minutes

Classroom Organization

Groups, followed by whole-class discussion

Materials

Out Numbered blackline master (transparency)

Doing the Activity

This activity requires little introduction. Emphasize that the focus of this activity, rather than "finding the answers," is to engage in discussion of the problems posed. Through such interaction, students will begin to create meaning for the relationships among graphs, tables, rules, and situations and, in part f of each question, how the concept of rate is connected to all the representations.

For example, if a group notices that a graph "goes up the same amount for each wagon," ask questions that might help the students notice how this amount appears, or should appear, in the table or rule. **How does the fact that 8 more people can be carried by each additional wagon affect your table? Can this number tell you how many people could be carried by 100 wagons? What would you do to calculate the number of people for *w* wagons?**

Discussing and Debriefing the Activity

When groups are ready, ask them to prepare transparencies to lead the discussion for parts d through f of each question. You may want to omit Question 1 to save time, as Questions 2 and 3 are more likely to need discussion. Also, you can begin the discussion before all groups have finished Question 3.

Use the presentations as an opportunity to connect the rules to the situations, which is the main idea in part f. Students may have noted that the numbers they found in answering Questions 1f and 2f showed up in the rules, and the discussion can emphasize the reason behind this. For example, in Question 1, students may see that the rule is something like *Out* = 8 • *In* and that there are 8 people per wagon. Ask why the number 8 appears both in the rule and as the number of people per wagon.

The geometric linearity of the graphs reflects the fact that each situation involves a constant rate. If you haven't yet used the phrase *constant rate*, now is a good time to work this language into the conversation. Encourage students to use the terms as well.

Question 2 involves a nonzero "starting value." The discussion of Question 2f should address this initial amount, which represents pounds of coffee on Day 0, as well as the rate.

Focus on how the rules offer a way to describe the situations, rather than on any formal procedure for getting such rules from In-Out tables. Although students may be able to find a rule by examining the table itself, help them to see the value of connecting the rule back to the situation. Some students may actually develop their rules directly by thinking through what's happening in the situation.

Introduce the synonymous terms *rectangular coordinate system* and *Cartesian coordinate system* for the standard graph setup with vertical and horizontal **axes** and equal-interval scales.

Remind the class that although their examples so far have involved only positive coordinates, axes are generally viewed as complete number lines, including negative as well as positive values. If you think it is needed, you might give students some coordinate pairs, including negative values, to plot.

Also be sure students know that unless otherwise indicated, *x* is used for the independent variable, which is represented on the horizontal axis, and *y* for the dependent variable, which is represented on the vertical axis.

Review the term **quadrant** and the standard numbering system as shown below. You might also remind students that the point where the axes meet, with coordinates (0, 0), is called the **origin**. Students will graph functions throughout their mathematics work, and you need not worry about whether they memorize the quadrant numbering system now. Casual references in context—for example, *Which quadrants does this graph use?*—will gradually familiarize them with the system.

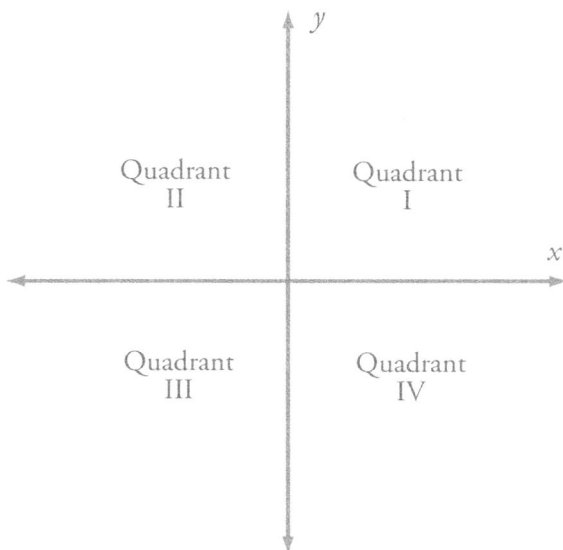

Use an equation from this activity to give meaning to the common phrase *graph of an equation*. For example, the graph in Question 2 is the graph of the equation $y = 25 - \frac{1}{2}x$.

Also mention that the set of all points that fit a rule is called the *graph of the equation* (or rule) and that the process of putting these points together to form an overall picture is called *graphing the equation*.

Key Questions

How does the fact that 8 more people can be carried by each additional wagon affect your table?

Can this number tell you how many people could be carried by 100 wagons?

What would you do to calculate the number of people for w wagons?

Why does the rule Out = 8 • In make sense in Question 1?

How much coffee is consumed each day in Question 2?

From Rules to Graphs

Intent

This activity reverses the process of moving from graph to rule followed in *Out Numbered*, helping students gain fluency interpreting representations and transferring between representations.

Mathematics

The multiple representations of a function—through situations, tables, graphs, and equations—is an important theme that will be revisited often in this unit and throughout the IMP curriculum. Later in this unit, the activity *All Four, One* provides an opportunity for students to synthesize their understanding of the connections among these representations.

In this activity, students create graphs for equations that are given symbolically. In the process, they are reminded that graphs can continue into quadrants other than the first, that a graph is like a picture of an In-Out table, that linear graphs have a constant rate of change (increase, decrease, or no change), and that not every graph is linear. The vocabulary and conventions of graphing are emphasized during discussion of the activity, encouraging students to use these terms to communicate their ideas.

Progression

Students were introduced to the phrase *graph of an equation* at the conclusion of *Out Numbered*. In that activity, students identified a rule from a graph; now they reverse the process by graphing given equations. The class discussion focuses on clarifying how students did their work, while reinforcing several mathematical ideas.

Approximate Time

5 minutes for introduction

25 minutes for activity (at home or in class)

20 minutes for discussion

Classroom Organization

Individuals, then small groups, followed by whole-class discussion

Doing the Activity

In *Out Numbered,* students started from graphs, made In-Out tables, and then found rules for those tables. Tell students that this activity reverses that process, and help them to clarify the instructions.

Discussing and Debriefing the Activity

Have students compare their graphs within their groups. Observe to see how students managed the instruction "sketch the graph." Ask, *How did you interpret the instruction "sketch the graph"? Do you think this graph will be discrete or continuous?*

Ask students from various groups to present their results. During the presentations, as students clarify and discuss their results, maintain the emphasis on students making sense of each other's work as you highlight key ideas, monitor the use of terminology and conventions, and review fundamental graphing concepts and terms introduced during the past few activities. Encourage use of vocabulary as appropriate, including *x*- and *y*-coordinate, ordered pair, *x*- and *y*-axis, coordinate system, plotting the point, quadrant, and rate.

One key idea to bring out is that the graphs involve more than just the first quadrant. In particular, for Question 1b, students need to use negative as well as positive inputs to see that the graph "goes up at both ends." You could introduce the word *parabola* for this shape; some students may recognize the term from previous math classes.

Help students see the connections between tables and graphs. *How is looking at a graph like looking at an In-Out table?*

Finally, focus attention on the graph for Question 1b to avert the potential misconception that all situations and all graphs are linear. Point out that the graph is not a straight line and connect this aspect of the graph with the behavior of the outputs in the table for this equation. For instance, the table might look like this (use integer inputs for simplicity).

In	Out
3	9
4	16
5	25
6	36

Ask how the fact that the graph is not a straight line is related to the values in the table. Students should see that, although the inputs are changing by the same amount at each step, the outputs are increasing by different amounts. They should gradually be coming to associate a constant change in outputs for a regular change in inputs with a linear graph.

Key Questions

How did you interpret the instruction "sketch the graph"?

Do you think this graph will be discrete or continuous?

Why do some rules give linear graphs that go up to the right while others give graphs that go down to the right?

How is the fact that the graph is not a straight line related to the table?

POW 7: Around the Horn

Intent

This POW, like others, is designed to engage students in exploring a larger problem over time. Writing about the problem—discussing the process of their investigation as well as justifying their solutions—is an important component of the problem-solving activity.

Mathematics

Ideas associated with constant rate are in play in this activity. In addition, students often demonstrate creativity and ingenuity in employing spatial and relational skills to model the proposed situation.

Progression

Students will be introduced to the problem and will begin to explore it in class. After a few days, if time allows, let them share ideas and generate methods for further exploration at home. After approximately one week, have a few students present the POW as a concluding activity.

Approximate Time

10 minutes for introduction

1 to 3 hours for activity (at home)

20 minutes for presentations

Classroom Organization

Individuals, followed by whole-class presentations

Doing the Activity

Have students read the problem statement, and give them some class time to begin working on it. The problem may seem deceptively simple at first. You may need to offer suggestions for how students might begin their investigation. Acting out the situation can be helpful. The actual physical movement of passing another person "midmonth" often generates significant insight for students. You might also suggest creating diagrams or using objects to represent the traveling ships.

Questions will probably arise about whether to count a ship that arrives in New York as your ship leaves or a ship that leaves San Francisco as yours arrives. The time difference between New York and San Francisco may also lead to a need for clarification. The manner in which these questions are resolved is not as important as students recognizing the need for making decisions of interpretation and clarifying their assumptions. Emphasize that students should make their assumptions clear in their write-ups.

This is a good POW on which to spend a few minutes of class time midway between the day it is introduced and the due date, encouraging students to share methods and findings. Students will often find that their peers have a variety of ideas, which can stimulate further thinking. If you don't have class time for this, at least check in with students to remind them to be working. Get a sense of how much progress they have made, and remind them of the importance of stating their assumptions in their write-ups.

Select three presenters for the discussion of the POW.

Discussing and Debriefing the Activity

Have the three selected students present their ideas about the POW.

There may be disagreement about whether to count the ships that are arriving as a student ship leaves and those that are leaving as the student ship arrives. Encourage classmates to ask the presenters to state their assumptions.

If any students created extensions of this POW, such as having ships leave once a day instead of once a month, you might have them present their work as well.

Key Question

What assumptions did you make?

You're the Storyteller: From Rules to Situations

Intent

This activity brings to a close the emphasis in *The Graph Tells a Story* on developing connections among the four representations of a function—rules, situations, graphs, and tables. It also continues the focus on the meaning of variables and algebraic expressions.

Mathematics

Students will draw upon the operations and numbers defining the relationship between two unknown quantities to create meaningful situations that given equations could define. They will also find solutions, via any method at their disposal, to make the equations true.

Progression

This activity works well as a homework assignment, followed by discussion and sharing of ideas. Later in the unit, students will summarize the connections among the four representations of a function.

Approximate Time

5 minutes for introduction

20 minutes for activity (at home or in class)

10 minutes for discussion

Classroom Organization

Individuals, followed by small-group or whole-class discussion

Doing the Activity

When introducing the activity, you might complete a simple example as a class to help students understand what is being asked of them. For instance, ask students to think of situations that would lead to the equation $2x = 8$. If needed, offer some examples, such as these.

- *Problem:* Donna studied for twice as long as Jerod. Donna studied for 8 hours. How long did Jerod study? *Variable:* Let x represent the number of hours Jerod studied. *Solution:* Jared studied for 4 hours.

- *Problem:* Sven is twice as old as his sister Tanda. Sven is 8 years old. How old is Tanda? *Variable:* Let x represent Tanda's age. *Solution:* Tanda is 4 years old.

Discussing and Debriefing the Activity

For the discussion, you might assign an equation to each group and have each group prepare a presentation of the situation that one of the members created for that equation. Or have each group choose a favorite example to share.

Encourage presenters, in addition to explaining how their situations fit the equations, to share how they came up with their ideas for the problems.

Key Question

What situation might lead to the equation $2x = 8$?

Traveling at a Constant Rate

Intent

The activities in *Traveling at a Constant Rate* are designed to deepen students' understanding of linear functions, including their representations and their use in modeling data and solving problems.

Mathematics

Linear functions are the specific focus of the activities in *Traveling at a Constant Rate*. Students fit lines to data using informal graphical methods (both by hand and with the support of technology) and use these lines to make predictions. They find symbolic rules for the lines and attend to two key features of linear functions—the *starting value* and the *constant rate of change*—and how these features are represented in graphs, tables, and rules.

Many of us have been taught that understanding linear functions means writing equations in a particular form. We may have a tendency to want to give students the abstract forms of equations of lines because that is our own experience. Rather than separating the symbolic work of algebra from reasoning within meaningful contexts, however, the intent in IMP is to give students extensive experiences with situations involving constant rate of change and finding linear functions for particular situations.

A major goal of the IMP curriculum is to provide students with contexts for developing mathematical concepts intuitively. Rate of change is a much bigger idea than slope of a line; slope is one specific case of the rate of change of a function. Looking at a variety of situations involving a constant rate of change gives students the opportunity to build a foundation for understanding the concept of slope. Through their work in *Traveling at a Constant Rate*, students will be able to write linear functions using situations and see the relationships among the rule, the table of values, and the graph.

Progression

Traveling at a Constant Rate begins with a set of activities that ask students to use given data to make predictions. The calculator is introduced as a graphing tool. As the activities unfold, students are asked to find linear rules and connect the numbers in these rules to features of graphs and tables, as well as to the context of the unit. In addition, students will present their results for the second POW of the unit and begin work on the third and final one.

Previous Travelers

Broken Promises

Sublette's Cutoff

Who Will Make It?

The Basic Student Budget

Following Families on the Trail

Graphing Calculator In-Outs

Fort Hall Businesses

Sublette's Cutoff Revisited

The Basic Student Budget Revisited

POW 8: On Your Own

All Four, One

Travel on the Trail

About James Beckworth

Moving Along

All Four, One—Linear Functions

Straight-Line Reflections

Previous Travelers

Intent

Students bring together many of the graphing techniques they have developed to plot data presented in tabular form and to analyze the data to make predictions. They assign a **line of best fit** to graphed data and then read and interpret new points along this line in order to make predictions and to identify a rule for the line. The intention is for students to recognize the graph of a function as the set of all points that satisfy the conditions of the function.

Mathematics

In this activity, students will

- transform information from tabular form or verbal description to graphical form
- look for a straight line that reasonably approximates the graph
- use a table for that straight line as an aid in finding a rule for the graph
- use the graph or the rule to make predictions and estimates about the situation

Although a linear model makes sense for this situation, the data do not perfectly fit a linear equation. Therefore, students will need to estimate and use common sense. Through this activity, they will be introduced to the concept of a line of best fit.

Progression

After the class does an example together, students work collaboratively to plot points, assign a line of best fit, and estimate a rule for that line.

Approximate Time

40 minutes

Classroom Organization

Groups

Materials

Clear straightedges or uncooked spaghetti

Doing the Activity

The class might first look at the map to place this section of the trip in geographical context. Fort Hall is in present-day Pocatello, Idaho.

Ask the class to read through the opening paragraphs and examine the table. Explain that they will be using this information to plan for groups of different sizes. You might explore the first set of data as a class. Ask, **How might you use the data about beans to plan for 20 people?**

One option students might suggest is to simply double the amount used for a group of 10 people or, similarly, to multiply the amount for 5 people by 4. When this idea comes up, ask what dilemma arises from this approach. Bring out that there are three different families with 10 people and that each of these three families used a different amount of beans.

Try to help students see the ambiguity of the situation and lead them to seek a way to use more of the available information. Students are likely to suggest graphing the data. Ask for the initial steps to be taken. **What do we need to do to make a graph for beans?**

Some students may not yet realize that certain steps must occur before others. Demonstrate their instructions, allowing any errors to become apparent. For example, students might begin plotting points without giving much thought to the scale. Start over as needed, which communicates the idea that when doing mathematics, the first approach may not turn out to be the best.

Students should understand that the multicolumn table in the letter can be viewed as shorthand for three separate In-Out tables, one for each item. If needed, create a separate In-Out table for the beans data.

Number of people	Number of pounds of beans
5	61
8	95
6	56
etc.	

The table contains 15 pieces of information. Use the term *data points* to refer to these pieces of information. Explain that each pair of numbers, (5, 61) for example, can be thought of as a coordinate pair and thus can be represented by a point in a coordinate system, hence the term *data point*.

If disagreement regarding which axis to use for which quantity arises, remind students that generally the *In* is associated with the *x*-axis and the *Out* with the *y*-axis. Point out that the amount of beans needed *depends on* how many people there are in a group. Thus "pounds of beans" is the dependent variable and goes on the vertical axis, while "number of people" is the independent variable and goes on the horizontal axis.

Now ask, **How should we scale the axes?** Students are asked to make predictions for the number of people in their own families, so they should ensure that the horizontal scale goes up to the largest family size they need. **Would you be able to use this scale to predict the amount needed for your family?**

Suggest starting both axes at zero. As discussed in *The Issues Involved*, starting an axis at a point other than zero can lead to distortion of the information. (It can be worth following up later about how starting at zero may have helped students determine their rules.)

Plot the points, and then solicit some observations about the visual representation of the data. Follow up these observations with the question, **What does the upward trend of the data tell you?** Students may notice that this "trend" is not uniform. However, the trend can be represented with a line. Tell students that such a line is called a **line of best fit**.

Ask a volunteer to place a pencil, a strand of dry spaghetti, or a clear straightedge across the data on the projector, or do so yourself. Have the class say when to stop so that the "line" is in the best position to represent the data. Keep the decision about the placement of the line intuitive, and acknowledge that the question of the "best" placement is subjective one.

Draw the line of best fit, extending it far enough that the amount of beans for each family can be found. Ask groups to read from the graph the number of pounds of beans that their Overland Trail families will need. **Based on the line of best fit, how many pounds of beans will each of your Overland Trail families need?**

You might also have students consider the special case of a family of size 0. Acknowledge that this isn't a family at all, but point out that students know for sure how many pounds of beans such a family would need—namely, no beans at all. Adding this point to the graph, and restricting the line of best fit to go through this point, makes the rule easier to find.

Next, turn students' attention to finding a rule for the line of best fit. Create a new In-Out table by listing the coordinates of several points spread out along the line. Then ask, **Can you write an algebraic rule for our line of best fit?** Many students may not understand the question; if the line on the graph has little meaning, points along the line won't either. Some students have yet to realize that the line represents many data pairs. Pay attention to developing these understandings over the next several activities.

Students' rule will be something like

number of pounds of beans = 12 • (*number of people*)

Point out that this equation can be used to find the pounds of beans for any number of people, including those not shown on the graph. First ask students to use the rule to figure the amounts for their own families. Then ask them to find the amount needed for their group's four combined families or for the class wagon train as a whole.

Instruct groups to tackle the other two supply items: sugar and gunpowder. You might give groups grid transparencies for creating their graphs. Save the graph of the beans data for use in the concluding discussion.

Discussing and Debriefing the Activity

In the discussion of this activity, emphasize the idea of the line of best fit and, once again, the relationships among graphs, tables, rules, and situations.

Ask groups to share their graphs as well as their thinking processes in developing the rule for the line of best fit.

At this point in the curriculum, rather than getting into a technical discussion of **slope**, help to begin to develop students' intuitive understanding of the concept. Slope is an abstract geometric concept related to, but distinct from, the concept of rate in a problem setting.

The data sets in *Previous Travelers* do not represent functions. You might raise this point by bringing out that the In-Out tables for the three items cannot represent functions. For example, there are two families with five members each, and each family uses different amounts of gunpowder, sugar, and beans. So the amount used for any of these commodities is not "a function of" family size. What students have done in this activity is to find graphs that give a reasonable approximation of the relationship between the amount used and family size. (Keep in mind, however, that a line of best fit *does* represent a function, except in the extreme case in which the line of best fit is vertical.)

Key Questions

How might you use the data about beans to plan for 20 people?

Will this technique yield the same prediction for 20 people if applied to the data for groups of 5 people?

How can you use all the data about beans to make a prediction?

How would you scale the axes?

Would you be able to use the scale you have assigned to this graph to predict the amount for your family?

How many pounds of beans would you estimate each person will use?

Based on the line of best fit, how many pounds of beans will each of your families need?

Can you write an algebraic rule for your line of best fit?

Broken Promises

Intent

Students practice plotting data and then using the graph to make predictions. In the process, they discover that in some situations, a linear approximation is not very useful.

Mathematics

Students create a set of data by estimating areas. They plot the data and then interpret and make predictions from the resulting graph. That the data involve unusual spacing and very large numbers challenges students to be thoughtful about the scales they select. That the data don't lend themselves to a linear model will encourage students to consider when and why to use a line of best fit.

Progression

Students work in their groups to help one another complete the process of plotting and interpreting the data, followed by a class discussion of the nonlinearity of the data.

Approximate Time

20 minutes for activity (at home or in class)

15 minutes for discussion

Classroom Organization

Individuals, followed by whole-class discussion

Materials

Broken Promises blackline master (transparency)

Doing the Activity

Little introduction to the activity will be necessary. Students may need to be reminded of how to measure area. You may want to point out the suggestion in Question 1 that students trace each map onto grid paper and use one square as the unit of area. Students will have difficulty estimating the shaded areas of the maps: you may want to mention that rough estimates will suffice.

Discussing and Debriefing the Activity

Let students compare results—especially their graphs for Question 2—in their groups, and then bring students together as a whole class. You may want to take a few minutes to talk about students' techniques for estimating the areas and for setting up the scales on the axes.

Then turn to Questions 3 and 4. **What are your predictions for the future of Native American land?** Based on the graph, one might expect all Native American land to have been gone by early in the twentieth century. In fact, though, there was comparatively little net change in the amount of land possessed by Native Americans during the twentieth century.

One conclusion to draw from this activity is that looking at a graph out of context can be fairly meaningless in predicting real-world events. A related conclusion is that using a linear approximation, or a line of best fit, is not particularly valid unless there is a good reason to expect the data to follow a linear pattern. You might connect the nonlinear nature of these data with the idea that the *rate* at which the Native Americans lost land was not constant.

Key Questions

How did you estimate the areas?

What are your predictions for the future of Native American land?

Supplemental Activity

Movin' West (extension) deals with questions about migration patterns in the United States and whether one can expect such patterns to continue. In addition to developing a general algebraic formula for a rate-of-change situation, students must think about comparative rates of change and do some commonsense reasoning to decide how well the model suggested by their formula might work in the future.

Sublette's Cutoff

Intent

Students make predictions and estimates based on limited data. This begins the development of their awareness of the connections between rate and starting point and the equation of a linear graph. The activity also gives students their first, informal experience reasoning about a system of linear functions.

Mathematics

Students continue to develop awareness of and flexibility with connections among situations, tables, and graphs. Though no rules are asked for per se in this activity, students will make predictions from the data, possibly employing some sort of curve-fitting technique. They also interpret graphs to answer informal, context-based questions related to intercepts, intersections, and rate.

Progression

After some time working in their groups, students share responses in a class discussion.

Approximate Time

30 minutes

Classroom Organization

Groups, followed by whole-class discussion

Materials

Colored pencils

Doing the Activity

Sublette's Cutoff started just west of South Pass, which was considered the halfway point on the trip to California. Point out on a map where the wagons are on the journey. South Pass is near present-day Highway 28, south of Lander, Wyoming, along the Continental Divide. Introduce the decision to be made at the cutoff: on the one hand, saving time, and on the other hand, the danger of little water or grass. Tell students that they will be given some information about the water supplies of three families who traveled the cutoff and will be asked to predict who will have enough water to complete the journey safely.

As students work, continue to assess how well they are graphing data and interpreting their graphs and to monitor the productiveness of group interactions. This activity lends itself to separate work, but students should also stay connected to what their group members are doing so that everyone is successful with the activity.

You might choose two or three groups—perhaps based on their use of alternative approaches—to give presentations. Select at least one group that settled on a linear model. Provide these groups with grid transparencies and pens. Remind them that they will talk about the key decisions they made in setting up their graphs and to be thoughtful about how they will explain the reasoning behind their answers for the other questions.

Discussing and Debriefing the Activity

If students bring up the unreliability of using the given data to make a prediction, have the class discuss the issue. Perhaps a student will make the point that people often need to make decisions when they do not have all the data and that an educated guess is better than a random decision.

One common error in setting up the scale is marking Day 2, Day 5, and Day 9 at equal intervals along the horizontal axis, forgetting to take into account that these time intervals are not all the same. Students may use a variety of approaches to answer Questions 2, 3, and 4. Linear and nonlinear models can be equally valid, provided students explain their estimates. Many students may reason numerically to answer these questions, rather than using the graph. This offers an opportunity to discuss the merits of each approach. Question 4 is part of the ongoing focus on rates and starting values and their connection with data and graphs. Encourage students to discuss what they had to do to arrive at their solutions.

Students might ask how long someone can go without water in dry conditions and while exerting great amounts of energy. Others may suggest that the three families share their water once one family's supply runs out. Encourage this sort of real-life discussion of the situation. It is important for students to see that mathematics is a tool that can help answer such questions, but does not necessarily tell the whole story.

Who Will Make It?

Intent

This activity gauges students' ability to make meaningful inferences from graphs.

Mathematics

Students examine another situation in which the data are approximately linear. They make estimates and predictions based on graphs of the data and linear models informally fit to these graphs. Students will also consider the meaning of the starting point as well as the downward trend in the data with regard to the situation and the graph.

Progression

Students will work on this activity individually, with an opportunity to check understanding with group members, and then compare and discuss methods as a class.

Approximate Time

5 minutes for introduction

20 minutes for activity (at home or in class)

20 minutes for discussion

Classroom Organization

Individuals, then groups, followed by whole-class discussion

Doing the Activity

The activity is similar to *Sublette's Cutoff* and requires little introduction.

Students may not think about using the fact that all three families started 330 miles from the Green River. Use your judgment about whether to suggest that they include (0, 330) as one of their data points or to leave this idea for the follow-up discussion.

Discussing and Debriefing the Activity

After students have had time to work individually, allow them to interact in their groups to first compare graphs and then to discuss answers and solution methods. You might select some groups to prepare transparencies for the discussion.

The discussion will likely follow similar lines as that for *Sublette's Cutoff.* Use any opportunities that arise to emphasize important vocabulary and conventions. In particular, encourage the use of the term **x-intercept** when students discuss methods for Question 2 and *constant rate* when discussing Question 5.

At some point bring out, if no one else does, that all three groups started 330 miles from the Green River and that students can use this information to provide an additional data point. Ask, **What location on the graph represents the starting point for each family?** They can use this point as part of their work in making predictions and estimations.

Also ask students to consider the downward trend in the data for each family. **How is the downward trend of the data evident in the graph?**

Key Questions

What location on the graph represents the starting point for each family?

How is the downward trend of the data evident in the graph?

The Basic Student Budget

Intent

This activity extends students' work with graphing and predicting to writing rules, setting the stage for building a contextually-meaningful understanding of symbolic representations of linear functions.

Mathematics

In a linear function of the form $f(x) = ax + b$, the y-intercept, b, may be thought of as the starting value (when considering first-quadrant data such as in applied contexts like those in this unit), and the **slope**, a, is the rate of change of y with respect to the change in x. This activity treats these concepts and connections informally.

Progression

Students will work in groups to plot data, sketch a line of best fit, and make predictions based on the line in order to answer questions about the data. The class discussion will involve some analysis of the numbers used in the rules and their association with the situation.

Approximate Time

25 minutes

Classroom Organization

Groups, followed by whole-class discussion

Doing the Activity

Tell students that this activity is similar to their recent work, but placed in a modern-day context.

As groups work, watch for opportunities to highlight students' use of the connections among the various representations of the situation—graphical, tabular, and symbolic.

If groups are having difficulty getting started, help them to see that one useful approach is to estimate the daily amount Cal, Bernie, and Doc each spend and to work from there. After making such an estimate, they are in a good position to find a rule. Essentially, what they need to do is to repeatedly subtract the amount each person uses each day from his starting amount. The formula that results is something like this, in which the value of x tells how many times to subtract the daily spending amount.

amount on day x = initial amount − (amount spent daily) • x

Keep in mind that "amount spent daily" is an estimate or average, as it is not the same each day.

Discussing and Debriefing the Activity

Questions 1 and 2 are similar to the prediction process used in *Sublette's Cutoff* and *Who Will Make It?*

Responses to Question 3 can be interesting. Some students will identify who cannot afford the concert; others may suggest they can share their money so all three can go. Some students may neglect to pay attention to having enough money at the end of the month for rent.

Focus attention in this discussion on Question 4—finding rules for Cal, Bernie, and Doc. Ask, **How did you find a rule for the amount of money each person would have?** This is more complicated than finding rules for many previous In-Out tables because no linear function fits the data perfectly.

Make sure students took into account the various starting amounts as well as the amounts in the table; there are four data points—not just three—for each roommate. For example, the information for Cal should include the amount of $1100 for March 31.

Ask students what connections they see between the rule and the situation. Some may identify the constant term as being equal to (or approximately) the amount of money the student began the month with. Some may identify the coefficient of the linear term as representing the amount spent, on average, per day.

If these observations are not forthcoming, you might draw them out by asking, **How can you see in the graph that Cal started with the most money? Can you see that in the rule as well?**

Address the idea of rate with the question, **Who spends the most money on average per day? Can you see this in the rule? How?** Someone may even recognize the nature of the subtraction (or negative rate) and its effect on the rule. **How does each student's graph tell you who is spending the most per day?**

Simply asking these questions to raise the awareness of starting values and rates is enough, as these ideas are more intentionally developed in subsequent activities.

Key Questions

How did you find a rule for the amount of money each person would have?

What connections do you see between the rule and the situation?

How can you see in the graph that Cal started with the most money? Can you see that in the rule as well?

Who spends the most money on average per day? Can you see this in the rule? How?

How does each student's graph tell you who is spending the most per day?

Following Families on the Trail

Intent

In this activity, students begin to study straight-line graphs in more depth. They identify connections between the coefficient in the symbolic form of a linear function and the related graph and situation.

Mathematics

Students' work shifts from fitting lines to data to working explicitly with a linear function of two variables. Students create and compare graphs involving constant rates, focusing on how the starting values and rates affect the graphs. They also create rules for constant-rate situations and examine how such rules depend on the rate and starting value.

Progression

Students work individually and then prepare in their groups to present methods and solutions to the whole class.

Approximate Time

5 minutes for introduction

20 minutes for activity (at home or in class)

20 minutes for discussion

Classroom Organization

Individuals, then groups, followed by whole-class discussion

Doing the Activity

Initiate the activity by having students read the scenario and Question 1a. Ask a few volunteers to summarize the information. Then ask for a volunteer to state what students are to do for Question 1a.

Tell students to rename July 12 as Day 0. This will eliminate differences in student approaches, which is not the emphasis in this activity.

Discussing and Debriefing the Activity

Ask each group to prepare a transparency of one of Questions 1 to 4. This preparation will provide a context for group members to answer one another's questions and to compare methods.

Begin with presentations for Questions 1a–c and 2a–c, focusing comments and inquiry on parts b and c by asking students to articulate how the starting values and rates affect the graphs.

For Question 1b, be sure students articulate that the Buck family graph "starts higher" than that for the Woods family.

For Question 1c, students will likely say that the graph for the Woods family is "going up faster" or something similar. Be sure to insert the word *rate* into this discussion.

For Question 2c, the discussion should make use of the term **parallel** to describe the two lines.

Ask the class, **Why do the graphs from Question 1 both rise as they go to the right while the graphs from Question 2 both fall as they go to the right?** Students should recognize that graphs that go "up to the right" represent situations in which the dependent variable increases as the independent variable increases, while the independent variable decreases for graphs that go "down to the right." The use of phrases such as "as time passes" rather than "as the independent variable increases" is appropriate while students are still making sense of these mathematical observations.

Have another member of the group present the group's responses to Questions 1d and 2d. Ask questions to heighten awareness of the connections among the situations, graphs, and rules. **How did the group use the information in the situation to write this rule? What information did the group use to write this rule? Is there information in the rule that shows up in the graph? How? Why?**

Questions 3 and 4 introduce the idea of using a pair of graphs to find a "common value." Solicit at least one presentation that demonstrates how a student used the graph to answer each question.

While the goal is to seed students' ideas about using a graph, discuss all methods used to answer these questions. For instance, in Question 3, someone might say that because the Woods family is gaining 5 miles per day on the Bucks and are 40 miles behind to start, they will catch up in 8 days.

For Question 4, students should see that the two families never have the same amount of coffee, because they are consuming coffee at the same rate. Make sure the graphical connection is made and uses the term *parallel lines*.

Key Questions

Why do the graphs from Question 1 both rise as they go to the right while the graphs from Question 2 both fall as they go to the right?

How did this group use the information in the situation to write this rule?

What information did this group use to write this rule?

Is there information in the rule that shows up in the graph? How?

Supplemental Activity

What We Needed (reinforcement) asks students to figure out how long it took for their families to travel from Ft. Laramie to Ft. Hall and how much of two commodities they would need to bring with them.

Graphing Calculator In-Outs

Intent

In this activity, students create and read graphs using the graphing calculator, in particular the $\boxed{\text{Y=}}$, $\boxed{\text{GRAPH}}$, $\boxed{\text{WINDOW}}$, and $\boxed{\text{TRACE}}$ keys.

Mathematics

Students employ technology to graph linear and nonlinear functions and use the graphs to answer questions. The trace feature of the calculator allows students to find In-Out pairs. The ability to use this technology is required for later investigations in the unit and throughout the IMP curriculum.

Progression

The teacher briefly introduces the basic steps for graphing a function on a graphing calculator before students work on the three parts of this activity in their groups.

Approximate Time

30 minutes

Classroom Organization

Groups, followed by whole-class discussion

Materials

Graphing calculators

Calculator manuals or guidebooks

Doing the Activity

The main goal of this activity is to acquaint students with the basic functions of their graphing calculators (or equivalent technology), including the following operations.

- Enter and edit functions and have the calculator draw their graphs

- Adjust the viewing rectangle, either directly or by using the Zoom feature, to see a particular portion of the graph and to get more precise information about the graph

- Use $\boxed{\text{TRACE}}$ to get the coordinates of a point on a graph

Some students may have discovered these operations on their own, but you should probably assume that most have had little experience with the graphing capabilities of their calculators.

Begin by illustrating these operations with one or two simple examples, starting with a linear function, such as an example from a recent activity. Then let students explore on their own for a while.

One point to emphasize is that the calculator will only graph an equation entered in the Y= form—that is, the form that expresses the output, which the calculator calls Y, directly in terms of the input, which the calculator calls X.

When students seem comfortable with these simple mechanics, have them begin the activity.

While students work, they are likely to struggle with several ideas. Encourage group members to work together to answer any questions. If an entire group has a question, limit what you show them to the Y=, WINDOW, ZOOM, GRAPH, and TRACE keys and associated menus on the TI-84 plus. Although the calculator provides several mechanisms for solving equations (such as 2nd [Calc 5:intersect and TABLE, do not demonstrate these at this time. The focus on the Trace feature is intentional, to continue to develop the connection between the visual location of, and meaning for, points on a graph in a coordinate system.

One of the mathematical challenges of this activity is understanding the need to adjust the viewing window in order to get the right part of the graph on the screen, as well as the steps for doing so. Finding a good viewing window is an essential skill for working with graphing calculators. Encourage students to play with the window limits and to develop their own strategies.

If a group is struggling to find a window they are satisfied with, help them to focus on the meaning of what they see in the graphing screen. One question you might pose is, What scales have you assigned to the calculator's display? What do you want to see? Or, What sort of numbers do you want to see that aren't in your viewing window? Are these in the x- or the y-direction? Once students make this decision, you can remind them how to adjust these settings in the WINDOW menu.

Another frustration students may encounter is being able to get better approximations than what the Trace feature initially provides, especially for Question 3. Remind them that they can use the Zoom feature or one of the options in the ZOOM menu. Often just this reminder is enough for students to figure out where to go next.

Discussing and Debriefing the Activity

Focus the discussion on how to read information from the graphs to answer the questions in the activity. This will likely entail some sharing of calculator-use methods and discussion of the Zoom and Trace features. Over time, students will grow more comfortable with manipulating viewing rectangles and other graphing-calculator techniques.

Key Questions

What scales have you assigned to the calculator's display? What do you want to see?

What sort of numbers do you want to see that aren't in your viewing window? Are these in the *x*- or the *y*-direction?

Supplemental Activity

Mystery Graph (reinforcement) is a graph-interpretation activity in which students are given the graph of a nonlinear function, but not its equation, and are asked to find a number of values for the function.

Fort Hall Businesses

Intent

In this activity, students examine linear situations in which they need to determine the starting value and the rate. Rather than developing a formalized method to derive rules from each type of situation, the goal in this activity is developing fluency in reasoning about linear functions and their representations.

Mathematics

Students use various kinds of information to find equations for linear functions. In one situation, they are given the rate and one data point and need to find the starting value in order to write the rule. In another situation, they are given two data points and must determine the rate, then the starting value, and finally the rule. In each case, they examine how the rule is related to the data and see the crucial role of the assumption that the rate is constant.

Progression

Students work individually and then compare results and methods in a class discussion.

Approximate Time

25 minutes for activity (at home or in class)

20 minutes for discussion

Classroom Organization

Individuals, followed by whole-class discussion

Doing the Activity

Let students know that they have arrived at their next milepost, and point out Fort Hall on the map. This activity begins by bringing up a choice many travelers faced: whether to continue on their journey or to make a home somewhere short of California. Many travelers did choose to settle along the trail.

Discussing and Debriefing the Activity

After students have worked individually on the activity, ask several groups to prepare presentations of either Question 1 or 2.

Have at least two presentations of each question, with at least one using a graph. Encourage the audience to ask questions as needed in order to understand how the presenters thought about the problem.

Ask the class, **How are these situations different from those in** *Following Families on the Trail?* For emphasis, follow with, **What didn't you know?** There are two key differences. Neither of these scenarios told students the starting value, and Question 2 did not explicitly give a rate. In the earlier problems, both values were given, which made finding a rule fairly straightforward.

Help students articulate the importance of the starting value in finding a rule for a situation. As students begin to appreciate its importance, they might give it a name of their own, such as *initial value*, *0-value*, or *y-intersect*. Introduce the term *y*-intercept and connect the idea of a **starting value**, where *x* is equal to zero, with the place where the graph hits the vertical axis.

In Question 1 students use the given information—$360 after 4 months and $50 per month—to find some other values. To find how much the Winstons started with they might use repeated subtraction or subtract four months' worth of deposits (4 · $50) all at once, to get $160.

For Question 1c, elicit various approaches. In a graphical approach, students might simply extend the line to where the *y*-value is $1000 and read down to find the *x*-value. In an algebraic approach, they could set up an equation, such as $160 + 50x = 1000$, and use a guess-and-check method to solve the equation. With a table, they could use a "month-by-month growth" method, repeatedly adding $50 to the account until the *Out* reaches, and exceeds, $1000.

If students get the exact, non-integer answer of 16.8, they will need to decide what to do. One approach is to point out that the Winstons make their deposits at the end of each month, meaning the account reaches and passes $1000 at the end of the seventeenth month.

For Question 2, be sure students clearly articulate how they found the ticket price. For instance, they might say they divided the change in the amount of money in the register by the number of tickets sold between the manager's two inquiries. Be sure they show the process with numeric examples as well as state the method.

Once students have found the individual ticket price, they might use either method described for Question 1, or some other approach, to get a rule. For example, if they use the fact that there was $70 in the register after the sale of 60 tickets, they will probably get the expression $70 + 0.75(p - 60)$. If they use the fact that there was $40 in the register after the sale of 20 tickets, they will probably get $40 + 0.75(p - 60)$. As with Question 1, use the variety of resulting rules to reinforce ideas about equivalent expressions and the distributive property.

If students see that the expressions in their rules can be simplified to $25 + 0.75p$, you could help them connect the starting amount and the individual ticket price to their roles in this simplified rule and in the graph.

Key Question

How are these situations different from those in Following Families on the Trail?

Supplemental Activities

Keeping Track, A Special Show, and *Keeping Track of Sugar* (reinforcement) offer additional opportunities for students to find starting values, rates of change, and rules for linear situations. All three involve constructing and using equations given two data points.

Sublette's Cutoff Revisited

Intent

In this activity, students use the calculator to replicate the pencil-and-paper process of determining a line of best fit.

Mathematics

Students reexamine the situation from *Sublette's Cutoff* using calculator-based curve-fitting techniques to make predictions. The emphasis is on developing the skills and techniques needed to use this technology. Students also see the graphical effects of changing the values of the linear and constant coefficients of a linear function.

Progression

Students learn how to use the calculator to plot points and then explore the results of adjusting linear functions to improve their fit.

Approximate Time

25 minutes

Classroom Organization

Whole-class discussion, followed by work in small groups

Materials

Graphing calculators

Doing the Activity

Remind students of the steps necessary to plot data and to sketch a line of best fit using pencil and paper. Then explain that the same work can be done using the calculator. They already know how to graph a line; ask them to restate the process learned in *Graphing Calculator In-Outs*. Today they will learn how to plot points.

Using the data from one of the families in *Sublette's Cutoff*, demonstrate how to enter and plot data points.

Once students have entered and plotted the points, ask how they might proceed to get a good function for the data. They should recognize that, previously, they sketched a line to match the data. If they wanted to know the equation of that line, they could determine it from the sketch. To use the calculator to graph a line, they must use the Y= menu.

Ask, **What might be a good function for these data points?** After an initial guess, students might prefer to adjust the linear term and/or the constant term of the function to improve the fit. Once they have done a little adjusting, pause to summarize the process and put them to work in their groups.

The basic calculator curve-fitting technique (it's still called that even when the function is assumed to be linear) is as follows.

- Plot the data.
- Leave the data on the screen and graph a function that you think might approximate the data well.
- Examine how closely the function's graph approximates the data. Adjust the function until you think it approximates the data as well as possible.

Have students plot the points for the other two families on the same graph and find functions that approximate each set of data. Group members can check that they get the same functions or, if not, investigate why they aren't the same.

Discussing and Debriefing the Activity

No debriefing of this activity will be necessary.

Key Question

What might be a good function for these data points?

The Basic Student Budget Revisited

Intent

Students practice the calculator method for curve fitting. The class discussion explores the effects on the graph of changing the coefficients of the linear function.

Mathematics

Students use the graphing calculator to fit a line to data and then use the line to make predictions. In the activity, students must use a negative linear term, which they connect to the context of the situation. They also identify the effects on the graph of $ax + b$ as a and b take on different values.

Progression

Students work individually and then come together to share results and to make connections between the symbolic form of the function, the situation, and the graph.

Approximate Time

5 minutes for introduction

25 minutes for activity (at home or in class)

15 minutes for discussion

Classroom Organization

Individuals, followed by whole-class discussion

Materials

Graphing calculators

Doing the Activity

Remind students of their work in *The Basic Student Budget*, and ask them to restate Cal's, Doc's, and Bernie's main concern. Some students might remember wanting to go to the concert; others will remember the need to have enough money at the end of the month for rent. Tell students that in this activity they will reexamine this question using their new graphing calculator techniques.

Discussing and Debriefing the Activity

The discussion can begin with volunteers showing what they have done so far in their work on this activity. After one or two presentations, give students an opportunity to confirm their own results in fitting a line to the data using the calculator, and then use the results to make predictions about who will be able to pay the rent.

Now direct students' attention to the symbolic form of the linear functions. Ask, **What is different about the lines in this graph compared to the lines in** *Sublette's Cutoff*? **What did you have to do in the equations to make the graphs turn downward?** Encourage students to justify why the subtraction, or negative, sign is necessary, based on the situation.

Take this exploration a bit further by selecting a starting value from one of the student presentations—representing the amount of money one of the students had at the beginning of April 1—and ask how students think this value affects the graph and the rule. Again, summarize student observations about the starting value and how it shows up in the graph and the rule.

Key Questions

What is different about the lines in this graph compared to the lines in *Sublette's Cutoff?*

What did you have to do in the equations to make the graphs turn downward?

POW 8: On Your Own

Intent

Students work on a research, planning, and budgeting problem involving thinking about their own future. This helps them to put mathematics to practical use and to develop their organizational skills.

Mathematics

To engage in budgeting and forecasting, the mathematical aspects of this activity, students must gather information and make decisions based upon those data.

Progression

Students are introduced to the activity and encouraged to begin to gather information immediately. Some class time midway through the project for students to share information and resources can be quite valuable. The project concludes with students sharing their lifestyle plans and budgets.

Approximate Time

10 minutes for introduction

1 to 3 hours for activity (at home)

15 minutes for discussion

Classroom Organization

Individuals, concluding with small-group discussions

Doing the Activity

This POW has a different flavor from others students have done, as it does not involve solving a mathematics problem. It does involve researching opportunities in the community and making decisions based on that research.

Introduce the activity of planning for life after high school and, in particular, developing a budget to support these plans. Have students read the activity on their own. Then ask them to summarize what is expected and to carefully identify the expectations and the structure of the write-up.

You may want to suggest that students work in pairs, perhaps pretending that they will be roommates sharing an apartment. Each student should prepare an individual write-up even if they collaborate on the POW. Also let students know that they will discuss this POW in their groups rather than making class presentations.

Get a progress report from students approximately midway to the due date. Because this is a different kind of assignment, students may need some guidance.

They can give each other advice about where to get information, and they might do further research via the Internet or by emailing questions to friends or relatives.

Discussing and Debriefing the Activity

In their groups, have students share what they learned. No specific mathematics needs to arise in these discussions. You might also allow the whole class to discuss what they learned from this activity.

Supplemental Activities

POW: High Low Differences is an additional open-ended POW in which students investigate sequences of calculations and look for patterns.

High Low Proofs (extension) asks students to prove conjectures they made in the *POW: High Low Differences* explorations.

All Four, One

Intent

This activity gives students a framework for reflecting on their work so far in the unit and will help you assess their understanding of the connections among the various ways to represent a relationship between two variables.

Mathematics

Students examine the connections among graphical, tabular, and symbolic representations of situations.

Progression

Working on their own, students reflect on the relationships among the four representations between two variables. The rest of the unit draws upon these representations for creating and manipulating linear equations.

Approximate Time

5 minutes for introduction

25 minutes for activity (at home or in class)

15 minutes for discussion

Classroom Organization

Individuals, followed by whole-class discussion

Materials

Students' work from previous activities

Doing the Activity

Take a moment to remind students that they have been studying relationships between two variables by considering graphs, tables, rules, and their associated situations. This activity is their chance to show what they know about these representations and how they relate to one another. Emphasize the expectations stated in the final paragraph of the activity.

Discussing and Debriefing the Activity

To initiate discussion, have volunteers read portions of their work aloud. Here are examples of the type of descriptive statements to look for.

- In-Out tables give specific points for a graph.

- A rule describes the relationship between the *Out* and the *In.*

- A graph is a visual representation of a situation.

- The coordinates of a point on the graph must fit the related rule.

Although you should take care to clear up misunderstandings, do not expect students to provide definitive statements about these relationships, which will be revisited many times in all four years of the IMP curriculum.

You may want to collect the responses to assess how well students understand the connections among situations, graphs, In-Out tables, and algebraic rules.

Travel on the Trail

Intent

Students consider how changes in starting value and rate affect graphs and rules.

Mathematics

This activity requires students to develop algebraic rules for given situations. They will use graphs or other representations of the resulting linear functions to answer questions about rate and solutions to a system of two linear equations.

Progression

Students will work collaboratively in their groups to put to use the connections they have made among situations, tables, graphs, and rules. Some class discussion will follow, primarily to emphasize the role of rate and starting value in each type of representation of the situations.

Approximate Time

30 minutes

Classroom Organization

Groups

Doing the Activity

Draw students into the activity by reading as a class the brief history of James Beckwourth in the student book. Groups that finish early can search classroom reference materials or the Internet to learn more about James Beckwourth.

The activity poses some mathematical challenges, beginning with interpreting the given information and determining how to structure the graph. Expect students to wrestle with these parts of the activity with their groups.

While monitoring, be prepared to help groups to interpret the graphing instructions in Question 1. Ask, **What clues do the instructions in Question 1a give you about the labels for the x- and y-axes?** Suggest that students assign Day 0 to July 28. If a group is stumped trying to draw the graph or write a rule, remind them that an In-Out table can be helpful.

For groups uncertain about how to approach Questions 2b or 2c, encourage them to draw a graph or a table so they can see more information about the situation.

It is important that everyone gets through Question 2b. You might have students who don't finish do so as homework or as an extension activity.

Discussing and Debriefing the Activity

There is no need to debrief Question 1. Request volunteers or have some groups prepare in advance to present Question 2, asking that they create a graph and write a rule as part of their response.

Focus the discussion of Question 2 on the connection between the numbers in the problem and the graph and rule that students develop. You can return to Question 1 to emphasize how and where student choices for rate and starting value affect the various representations.

During the discussion, it is useful to mention points on the graph and to record their coordinates using conventional notation—for example, (0, 23). If necessary, ask students to interpret the meaning of this notation.

If time allows, have the students in a group present their work on Question 3 to prompt further discussion. Because every group's response will be different, students will have to communicate carefully the methods they used to solve the problem, rather than simply stating the solution.

Key Questions

What clues do the instructions in Question 1a give you about the labels for the x- and y-axes?

How does the 12 show up on the graph? In the rule? Why?

How does the 23 show up on the graph? In the rule? Why?

Moving Along

Intent

This activity continues to engage students in solving problems involving four representations of linear functions. They are specifically asked to write equations and to identify rates and starting values, with the goal of strengthening connections between the situation and the rule for linear functions.

Mathematics

Students continue to draw upon their understanding of the relationships among the four views of a function in order to develop equations for situations in which two data pairs are provided. The roles of rate and starting value are emphasized, setting the stage for formalization of the standard form of a linear function.

Progression

After students work individually on the activity, they come together as a class to discuss methods for determining the rate and starting value and for writing the equation.

Approximate Time

5 minutes for introduction

20 minutes for activity (at home or in class)

15 minutes for discussion

Classroom Organization

Individuals, followed by whole-class discussion

Materials

Students' work on *Previous Travelers*, *Sublette's Cutoff,* and *Who Will Make It?*

Doing the Activity

This activity will be quite familiar to those who have taught a traditional algebra I course; students are determining the equation of a line given two points. The goal of this activity, however, is not simply for students to be able to write the equation of a line given two points. Rather, these scenarios are opportunities for students to reason about linear functions, drawing upon the connections among the four representations to identify processes that will help them write equations. With that in mind, encourage students to use what they know to help them to think about these scenarios rather than being concerned that they "discover" a tried-and-true shortcut procedure.

Discussing and Debriefing the Activity

Students may benefit from having a few minutes to share responses and methods in their groups. During this initial discussion phase, pass around a few transparencies for volunteers to record their methods.

Focus the discussion on clarifying the presenter's methods, including the mechanics and thinking behind getting the equations and how the numbers in the equation are related to the situation. Although students were not explicitly asked to make graphs of the lines, you may want to ask for them as part of the discussion.

In finding the equation for Question 1, students may have answered the first part of Question 1b (finding the rate) before Question 1a (getting the equation).

To get the rate, students might use just a single point and the common-sense approach of dividing the amount of beans by the number of people. For instance, the point (4, 48) represents needing 48 pounds of beans for 4 people, so each person needs 12 pounds. If a presenter demonstrates this approach, try to also identify someone who used both points, (4, 48) and (10, 120), and reasoned that the additional 6 people (the difference between 4 and 10) required an additional 72 pounds of beans (the difference between 48 and 120), which again gives 12 pounds per person.

Students will then likely intuitively understand that they can simply multiply the number of people by the amount per person to get a rule like $B = 12N$.

For the second part of Question 1b, students need only note that the number indicating the rate is the same as the coefficient of the independent variable.

Elicit a variety of approaches for Questions 2 and 3, emphasizing the connections to the contexts. A few students may recognize that the rate always shows up in the equation as the coefficient of the independent variable and work from that.

For Question 3, have one or two students present their contextual settings and describe what the two points mean in that context. Talk about how to develop the equation from the coordinates of the points and the contextual significance of the numbers in the equation.

To conclude this activity, ask groups to write out a general process—an algorithm—to determine the equation of a line if they are given two points.

All Four, One—Linear Functions

Intent

In this summative activity, students describe, in writing, methods for moving from one representation of a linear function to another.

Mathematics

This activity draws upon and strengthens the connections among the four representations of a function, with a particular focus on linear functions. Students are encouraged to work from rate of change and starting value in identifying these connections. They learn that *linear* has an algebraic as well as a geometric meaning and are introduced to a common form of the linear function, $y = ax + b$.

Progression

Students work on this activity in groups, perhaps submitting a first draft submitted for feedback, with a final draft due at a later date.

Approximate Time

30 minutes

Classroom Organization

Groups

Materials

Students' work on the activities in *Traveling at a Constant Rate*

Doing the Activity

Remind students that the term *linear* was initially used to refer to a function or situation that led to a straight-line graph. (See the introductory comments on *Wagon Train Sketches and Situations*.) Then ask, **How can you tell from looking at a rule whether its graph is a straight line?**

The goal here is for students to recognize that rules for straight lines have a specific algebraic form. Students might reply with something like $y = s + rx$, where s represents the starting value and r represents the rate, likely suggesting numbers in place of the variables s and r. Help students recognize that the s- and r-values can be positive, negative, or zero. (See the "Mathematics" section in the introduction to *Traveling at a Constant Rate* for a relevant discussion.)

Ask, **Why do situations with constant rates have rules of this special type?** Students may offer a variety of responses to this question.

Explain that rules of this type—namely, of the form $y = s + rx$—are called **linear functions** and that this term has both a geometric and an algebraic meaning.

Also explain that a common way to write a rule for a linear function is $y = ax + b$. Ask students to identify the meaning of the a and b terms. Also state that any rule equivalent to a rule of this form is linear as well.

As groups work, help students to recall relevant activities from the unit. Do you remember an activity from this unit in which you had to create an In-Out table from a graph?

Discussing and Debriefing the Activity

No formal debriefing of this activity is necessary.

Key Questions

How can you tell from looking at a rule whether its graph is a straight line?

Why do situations with constant rates have rules of this special type?

Do you remember an activity from this unit in which you had to create an In-Out table from a graph?

Straight-Line Reflections

Intent

This activity introduces the idea of equivalent forms of linear functions by drawing on students' ability to move among the various representations of a function.

Mathematics

Students draw upon their understanding of the four representations of linear functions to informally become acquainted with the notion of equivalent equations. In particular, they graph an equation and then use what they have learned to determine a linear equation of the form $y = ax + b$ from the graph. Students will then apply the distributive property as well as other basic symbol manipulation to demonstrate the equivalence of the two equations.

Progression

Students work individually on the activity and then debrief their work as a class, making observations and conjectures about various symbolic forms of equivalent linear equations.

Approximate Time

5 minutes for introduction

30 minutes for activity (at home or in class)

15 minutes for discussion

Classroom Organization

Individuals, followed by whole-class discussion

Doing the Activity

Introduce this activity as an individual investigation. (Note that the activity makes no attempt to define **equivalent equations**. A careful definition of equivalence—one that would involve sets of solutions—may detract from students bringing their own reasoning to the problems.)

Discussing and Debriefing the Activity

Discussion can begin immediately with student presentations of Question 3a. Encourage discussion to clarify the presenters' methods and reasoning. Also request presentations of Question 4a with the same goal—a discussion of a variety of approaches.

Then ask for ideas about how to approach Questions 3b and 4b. Ask, **What did you interpret the word *equivalent* to mean in Questions 3b and 4b? Where have you heard the term before?**

Students may observe that the ordered pairs that satisfy one equation also satisfy the other or that the related In-Out tables or graphs are the same. Others may say the equations are not equivalent because they don't appear to be the same. Tell students that when two equations are **equivalent**, every pair of numbers that satisfies one equation also satisfies the other. With this definition, the two equations in Question 3 are equivalent, as are the two equations in Question 4. Let students know that they will return to this idea later in the unit

Key Questions

What did you interpret the word equivalent to mean in Questions 3b and 4b?

Where have you heard the term before?

Supplemental Activities

The Growth of Westville (extension) provides a western setting for examining situations that may appear to involve constant growth but do not lead to linear graphs, so this is a good follow-up to the series of activities.

Westville Formulas (extension) is a follow-up to *The Growth of Westville*.

Reaching the Unknown

Intent

The activities in *Reaching the Unknown* support students' understanding of linear functions and their representations by emphasizing symbolic representations and connecting this work to that done in earlier activities.

Mathematics

The emphasis of these activities is on symbol sense. Students focus on symbolic representations of linear functions and on developing a meaningful understanding of some of the symbol-manipulation procedures used to solve equations. The context of the unit, the distributive property, and a mathematical model for symbol manipulation all help support these procedures. The context supports understanding of what the symbols and their relationships represent. The distributive property allows students to derive equivalent equations. The "mystery bags" model, a pan-balance model of equations, gives students a metaphor for symbol manipulation.

Flexible algebraic thinking includes the processes of "doing" and "undoing," such as the ability to find equations with given solutions as well as to find solutions to given equations. In these activities, students will work explicitly on developing this algebraic flexibility.

Progression

Students will answer a variety of questions by writing and solving symbolic equations and systems of equations. They will revisit previous questions solved graphically and solve them symbolically. In addition, they will present their results on the final POW of the unit, compile unit portfolios, and complete in-class and take-home unit assessments.

Fair Share on Chores

Fair Share for Hired Hands

More Fair Share on Chores

More Fair Share for Hired Hands

Water Conservation

The Big Buy

The California Experience

Getting the Gold

The Mystery Bags Game

More Mystery Bags

Scrambling Equations

More Scrambled Equations and Mystery Bags

Family Comparisons by Algebra

Starting Over in California

Beginning Portfolios

The Overland Trail Portfolio

Fair Share on Chores

Intent

In this activity, students write equations, express them in equivalent form, and graph them. Two important concepts are reinforced in the process: abstracting a problem to an algebraic equation and reading numeric information from a graph. Students convert a linear equation into Y= form, conducive to use on the graphing calculator. The activity sets the stage for students to make meaning of graphical solutions to systems of equations.

Mathematics

Students examine possible solutions to a linear equation that involves two variables and explore how to express that equation by giving one variable in terms of the other. They also solve equations for one variable in terms of the other, based on reasoning in the problem context. Finally, they graph linear conditions on the calculator and use this tool to find solutions to an equation.

Progression

Students will work on the activity in groups and then share their results with the class. This activity sets the stage for lending graphical meaning to solving systems of equations. Students will revisit this situation in *More Fair Share on Chores*.

Approximate Time

30 minutes

Classroom Organization

Groups, followed by whole-class discussion

Materials

Graphing calculators

Doing the Activity

Read the introduction as a class. If you have a map, you might point to the split between the California and Oregon Trails.

You may want to have a volunteer offer a sample solution for Question 1 as a way to get the class started. You may need to prompt students to consider values other than whole numbers.

Circulate as groups work. In Question 2, students are asked to express a problem condition by using an equation. Although this may seem straightforward, translating words into symbols can be a complex process. If an entire group is stumped, remind them of the strategy of identifying and writing in words the arithmetic used. In this case, it is likely that students are performing an arithmetic operation like "multiply the girls' hours by 2 and the boys' hours by 3, add the results, and see if the total equals 10."

If groups are stuck on Question 3 or 4, focus them on the arithmetic. **If each girl's shift was four hours, how would you find the length of each boy's shift?** Help them to translate their verbal reply into symbolic form—an equation like

$$B = \frac{10 - 2G}{3}.$$

Discussing and Debriefing the Activity

Begin the discussion by asking students to compile their possible answers to Question 1 into an In-Out table. Because students will be expressing the boys' shift length, B, in terms of the girls' shift length, G, the girls' shift length should go in the *In* column. Have students plot these points.

If students had trouble with Question 2, ask them to use a number pair from the table and express in words how it gives the desired total of ten hours. They will probably reply by saying something to this effect: "Multiply the length of a girl's shift (such as $\frac{1}{2}$) by 2 and the length of a boy's shift (such as 3) by 3 and add."

Essentially, they have given you the equation; they just need to restate it symbolically.

Help students to see that the purpose of Questions 3 and 4 is to express the equation in the form B= so they can graph it on the calculator. Ask, **Why might you want to express one variable in terms of the other? If you know the length of a girl's shift, how would you use it to find the length of a boy's shift?** Otherwise, they could stick with the equation $2G + 3B = 10$, which is a more natural way to think about the problem.

Here are some observations that might arise in the discussion.

- The longer the shift for each girl, the shorter the shift for each boy.

- There is a maximum possible shift length for each group.

- The points that fit the equation lie on a straight line.

- Any point that is on the graph fits the equation, and vice versa.

Post the equation from Question 4 so students can refer to it in the next activity.

Key Questions

If each girl's shift was four hours, how would you find the length of each boy's shift?

How can you check whether a number pair fits the condition?

Why might you want to express one variable in terms of the other?

Fair Share for Hired Hands

Intent

Students examine possible solutions to a linear equation in two variables and explore how to express that equation as one variable in terms of the other. The activity focuses on the connection between an equation and its graph to further set the stage for considering the graphical meaning of the solution to a system of equations.

Mathematics

Students continue work with all four representations of linear functions. Most importantly, they write a rule in Y= form and consider the meaning of specific points on the line.

Progression

Students use a context to derive a set of ordered pair solutions, plot these pairs, and determine an equation for the linear relationship. Again, students are asked to convert the equation into Y= form. Finally, a class discussion emphasizes the meaning of the points along the graph of the line. This sets the stage for the next activities, in which students consider the meaning of a point of intersection.

Approximate Time

5 minutes for introduction

20 minutes for activity (at home or in class)

15 minutes for discussion

Classroom Organization

Individuals, followed by whole-class discussion

Doing the Activity

Let students know that this activity is similar to the previous activity, *Fair Share on Chores.* Although students have graphing calculators available, suggest they initially draw the graph on paper.

Discussing and Debriefing the Activity

Have students compare ideas and answers in their groups. You might have several students put their solutions for Question 1 and 2 on the board. After a few minutes, gather the class and have some discussion of the responses.

Next, turn to Question 4, eliciting verbal descriptions of how to get Y from X, and comparing the different versions.

Discuss the equations students wrote for Question 5, which probably look much like $Y = \dfrac{20 - 3x}{4}$. As a class, test some of the points students found from their graphs in Question 3.

At this point you can ask the class, What is the relationship between fitting the equation and being on the graph? Students should come to this conclusion:

Every point on the graph represents a solution to the equation, and every solution to the equation corresponds to a point on the graph.

Essentially, this is the definition of a **graph**.

Then ask, Is every solution to the problem on the graph? Does every point on the graph represent a solution to the problem?

The discussion should bring out the fact that a point on the graph, which represents a solution to the equation, does not necessarily represent a solution to the problem. For example, the point represented by $X = 4$ and $Y = 2$ is on the graph and fits the equation, but is not a solution to the problem, because the experienced workers should have a higher pay rate than the inexperienced workers.

This tells us that even though the graph represents the equation, the equation itself does not precisely represent the problem. The graph for the actual problem is only a portion of the line. In other problems, only points with whole-number coordinates will fit the problem.

Key Questions

How did you get Y from X?

What is the relationship between fitting the equation and being on the graph?

Is every solution to the problem on the graph?

Does every point on the graph represent a solution to the problem?

More Fair Share on Chores

Intent

Students examine how to use graphs to find the solution that fits two linear conditions.

Mathematics

Students write and graph an equation representing a new linear condition for the situation first encountered in *Fair Share on Chores*. Next, they are asked to find a solution that fits both this new condition and the original condition; that is, they are asked to find the solution to a system of two linear equations in two unknowns (although that formal mathematical terminology is not used).

Progression

Students will work on the activity in groups and then share their solution methods with the class.

Approximate Time

30 minutes

Classroom Organization

Groups, followed by whole-class discussion

Materials

Students' work on *Fair Share on Chores*

Doing the Activity

Tell students that this activity is similar to the work they have been doing.

Students should not have much trouble with Question 1. If they do, help them recall the methods they have used recently.

Allow groups to struggle to find their own ways to solve Question 2. Here are three strategies they might use.

- They can graph the two equations on their calculators and use the Trace feature to find the coordinates of the point of intersection.

- They can graph just one of the equations and use the Trace feature to look for a point on the graph whose coordinates satisfy the other equation.

- They can compare the In-Out tables to see whether they have a common entry.

When each group has found a solution, pick two or three to present their methods to the class.

Discussing and Debriefing the Activity

Begin, if you think it is needed, with a discussion of the individual parts of Question 1, or just have a student give the equation from Question 1c, $B = G + \dfrac{1}{2}$.

Then let the chosen groups present their work on Question 2. If no group uses a graphical method to solve the system of equations, lead the class through a graphical solution. Ask, **How did you represent graphically the pairs of shift lengths in which each boy's shift is half an hour longer than each girl's shift?**

How did you represent graphically the pairs of shift lengths that total ten hours?

Students should recognize that each condition, by itself, is represented graphically by a line. If necessary, emphasize that every point on a given line fits the equation for that line. It shouldn't be hard to get someone to articulate that we want to find a point that is on both lines.

Because the point of intersection is a point on both graphs, it must satisfy both conditions of the problem. Ask several students to state this observation in their own words.

Since students have both equations in a form that gives B in terms of G, they can draw the two graphs simultaneously. They can use the Trace feature to estimate the coordinates of the point of intersection. The point is exactly (1.7, 2.2), which translates to 1 hour, 42 minutes for each girl and 2 hours, 12 minutes for each boy. Have students verify that this pair of times fits both conditions of the problem.

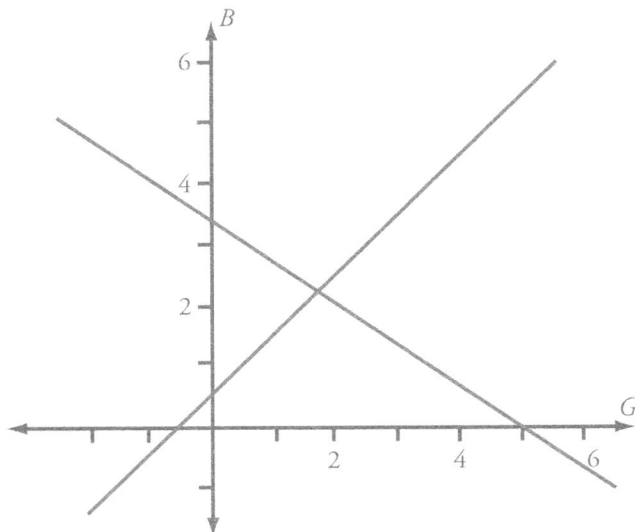

Key Questions

How did you represent the first condition graphically?

How did you represent the second condition graphically?

What points fit both conditions?

More Fair Share for Hired Hands

Intent

This activity, a follow-up to *More Fair Share on Chores* that builds on students' work on *Fair Share for Hired Hands*, will provide information on students' understanding of the connection between graphs and equations.

Mathematics

Students write and graph an equation for a linear situation. Next they determine a solution to a pair of conditions represented by linear equations and are asked to find the solution to a system of two linear equations in two unknowns.

Progression

Students will work on this activity individually and then share their techniques as a class. The activity concludes with a discussion of graphical methods that emphasizes that every point on a line fits the equation for that line.

Approximate Time

5 minutes for introduction

20 minutes for activity (at home or in class)

15 minutes for discussion

Classroom Organization

Individuals, followed by whole-class discussion

Materials

Students' work on *Fair Share for Hired Hands*

Doing the Activity

Tell students that this activity extends *Fair Share for Hired Hands* in the same way that *More Fair Share on Chores* extended *Fair Share on Chores*. To encourage students to use a graphing approach, you might suggest they use the graph from *Fair Share for Hired Hands* to begin this work.

Discussing and Debriefing the Activity

Begin the discussion by having students compare results and graphs in their groups.

As before, there are several ways to approach the problem, and students may not have solved it by graphing the two equations. They should have the graph of the equation $Y = X + 1$ for Question 2, but they might have answered Question 3 by a guess-and-check method.

Observe the group interactions to gauge how well students can plot graphs and write rules and to get a sense of how well they understand that the points on a graph are solutions to the equation or to the situation that the graph represents.

Ask groups for an equation that represents the additional information provided in Question 3. If they come up with $3X + 4Y = 30$, ask how they would put this equation in a form that would allow them to graph it on the calculator. You may want to suggest that they think about how they would find the rate for each more-experienced hand if they knew the rate for each less-experienced hand.

If needed, have them go through the arithmetic with a specific value for X and then analyze the arithmetic steps (in a manner like that described for *Fair Share on Chores*). They should be able to express the condition that the total pay is $30 in a form similar to $Y = \dfrac{30 - 30X}{4}$.

Have all students plot both graphs on the calculator, use the TRACE key to locate the point of intersection, and confirm that the solution they found earlier matches this work. Use this work to help emphasize that points on the graph are solutions to the equation and that the intersection point is a solution to both equations. Ask students to articulate how they know that the point of intersection yields the solution for both conditions of this problem. **How might you convince someone that the coordinates of the point where these two lines intersect tells where both conditions in this problem are satisfied?** Their replies might include the notion that if you check that point in each equation or in each situation, it makes a true statement.

Key Questions

What equation represents the information in Question 3?

How can you rewrite the equation $3X + 4Y = 30$ so you can graph it on the calculator?

How would you find the rate of pay for each more-experienced hand if you knew the rate for each less-experienced hand?

How might you convince someone that the coordinates of the point where these two lines intersect tells where both conditions in this problem are satisfied?

Water Conservation

Intent

This activity gives students more experience in writing rules for situations and in finding the common solution to a pair of equations.

Mathematics

Students continue their exploration of linear functions. In particular, they examine ways to find the input for which two linear functions give the same output.

Progression

Students work on the activity in groups and then discuss their work as a class. The equation developed by the class for Question 5 will be returned to later in the unit when students have the symbol-manipulation tools necessary to solve it algebraically.

Approximate Time

30 minutes

Classroom Organization

Groups, followed by whole-class discussion

Doing the Activity

Tell students that paying attention to water supplies was an important aspect of life in the Overland Trail era, as it still is today. This activity explores some questions about monitoring water supplies. Point out the setting, northeast Nevada, on the map.

As students work, give various groups transparencies to use in preparing for the class discussion.

Discussing and Debriefing the Activity

Most of this activity's content should be familiar to students. This discussion will serve as the context for bringing out some new ideas. After presentations of Questions 1 and 2, record the expressions students developed ($50 - 3X$ and $100 - 8X$) for use following the discussion of Questions 4 and 5.

If students have difficulty with Question 4, remind them that one of the lines represents the amount of water the Stevens family has on a given day and the other line represents the amount of water the Muster family has on a given day. They should recognize that to find a day when the two families have the same amount of water, they want to look for a point that is on both lines.

Ask how the answers to Question 5 show up on the graph. Bring out that the solutions are the *x*-coordinates of the points where the graphs cross the *x*-axis. Introduce the idea of **x-intercept** as being the point on the graph where the *y*-value is zero.

Now is a good opportunity to set the stage for solving these types of problems algebraically. Ask how students might use their expressions from Questions 1 and 2 to write a single equation that could be used to identify the day when the families have equal amounts of water. If needed, emphasize that the two expressions represent the amount of water that the two families have after *X* days, so for the families to have the same amount of water, the two expressions must represent the same value.

The idea of using these expressions to create an equation is often a conceptual leap for students. You might use a guess-and-check approach to help them see this connection. For instance, ask whether the families had the same amount of water after six days and how they know. Students should see that they can answer this by evaluating the expressions $50 - 3X$ and $100 - 8X$ at $X = 6$ and noting that the results are different. For the families to have the same amount of water, the results must be the same—that is, $50 - 3X$ and $100 - 8X$ must be equal for a specific value of *X*.

Students should have noted from the graph that the lines intersect at (10, 20). Ask what this means about the equation $50 - 3X = 100 - 8X$. Bring out that when *X* is 10 (the first coordinate), both sides of the equation equal 20 (the second coordinate).

Post the equation $50 - 3X = 100 - 8X$ for further discussion after *More Mystery Bags*.

Key Questions

How can you use your expressions from Questions 1 and 2 to get an equation for finding the day when the families have *equal* amounts of water?

Supplemental Activity

The Perils of Pauline (extension) is a well-known but challenging puzzle problem. Students are given information about the speed of an oncoming train and the position of a person in a tunnel that the train is approaching and are asked to determine the person's speed given that she made it out of the tunnel on time.

The Big Buy

Intent

This activity continues, in a present-day context, students' work writing equations to represent linear conditions and then graphing to solve for given sets of conditions.

Mathematics

Students continue their work with linear functions and use a pair of graphs to compare different rates of pay. In particular, they graph to find the input for which two linear functions give the same output. The concept of the point of intersection plays an important role in this problem. The activity concludes with students writing an equation whose solution would solve the problem.

Although students continue to solve simple systems of equations in this activity, no formal work is done to name this process or to reduce it to a rote procedure.

Progression

Students work individually to develop and graph two equations for a given situation and then to answer questions about the situation. They then share ideas in their groups and also write an equation that determines the x-coordinate of the point of intersection of the two lines.

Approximate Time

20 minutes for activity (at home or in class)

10 minutes for discussion

Classroom Organization

Individuals, followed by small-group and whole-class discussion

Doing the Activity

This activity requires little or no introduction.

Discussing and Debriefing the Activity

Have students share answers, check each other's solutions, and answer questions in their groups.

This activity has students go straight to writing equations without suggesting that they look at specific cases or make a table. If necessary, remind them that an In-Out table is always a good tool to use if they are having difficulty understanding a situation.

Circulate as groups compare results to see whether any class discussion is warranted.

As a follow-up to Question 5, ask students to think privately about this question: **What equation would tell us how many hours of work it would take for Jillian and Max to earn an equal amount of money?** If necessary, remind them of the similar question posed during discussion of *Water Conservation.* Students should see that they can set the expressions from their equations in Question 1 equal to each other, giving the equation $10 + 5x = 40 + 3x$. Verify that students' answer for Question 5 fits this equation, and bring out that each side of the equation gives the common amount that Jillian and Max would earn.

Post the equation, $10 + 5x = 40 + 3x$, for discussion after *More Mystery Bags*.

Key Question

What equation would tell us how many hours of work it would take for Jillian and Max to earn an equal amount of money?

The California Experience

Intent

Students read this reference page for historical background, learning that the great migration of the mid-nineteenth century was not a positive experience for all who undertook it.

Mathematics

This reference page continues the story of the history of travel along the Overland Trail.

Progression

Students read the material as a class, followed by an opportunity to comment and ask questions. The storyline sets the stage for the next activity, *Getting the Gold.*

Approximate Time

10 minutes

Classroom Organization

Whole class

Getting the Gold

Intent

In this activity, students extend their work with linear functions beyond the first quadrant, giving meaning to the *x*-intercept of a graph. This activity also generates one more equation obtained by setting two linear expressions equal, setting the stage for the development of methods for solving simple linear equations.

Mathematics

Students continue their work with representing situations by equations. In this case, they consider the starting values and rates related to the profit a miner might make using two different methods for collecting gold. They compare the associated linear functions both graphically and symbolically, considering questions that can be answered by the *x*-intercepts and the point of intersection or by setting up an equation for finding when the two functions are equal.

Progression

Students work individually to explore two methods for collecting gold and the profit associated with each, using graphs to answer questions about break-even and equal-profit points. In a class debriefing, they develop and solve linear equations to answer the same questions.

Approximate Time

5 minutes for introduction

20 minutes for activity (at home or in class)

15 minutes for discussion

Classroom Organization

Individuals, then groups, followed by whole-class discussion

Doing the Activity

After students read *The California Experience* to put the gold rush in perspective, introduce them to this activity. Make sure they understand that profit is income minus expenses. If the activity is assigned as homework, it is a good idea for students to start in class, even briefly, to ensure they know how to approach Question 1.

Discussing and Debriefing the Activity

Have students debrief their work in their groups. Invite a few students to post their expressions from Question 3a and 3b on the board, as these will be used in discussing Questions 4 and 5. Also ask a group or two to prepare transparencies to lead discussion of Questions 4 and 5. One group's solution for each will be sufficient, especially if students are becoming proficient with questions like these.

After Questions 4 and 5 are presented, review the graphical and algebraic ways to approach these types of questions.

First, note that the starting values are both negative and give the places where the graphs cross the y-axis.

Then point out that the break-even point is the value where each graph crosses the x-axis. Bring out that finding the break-even point is like solving an equation in which the expression for profit is set equal to 0. Thus, the panning method breaks even at the point where the value of x gives a solution to the equation $15x - 60 = 0$.

Finally, emphasize once again that finding the equal-profit point is the same as finding the point of intersection of the two graphs and that this point can be determined by solving the equation formed by setting the two algebraic expressions equal to each other. By now, this will be routine thinking for some students. You may want to pose it as a question for students who need to give this relationship additional thought.

Students can verify that the x-coordinate of the point of intersection on the graph fits the equation $15x - 60 = 30x - 420$. You can also bring out that the y-coordinate of this point is the value of each side of this equation.

Post this equation, $15x - 60 = 30x - 420$, for discussion after *More Mystery Bags.*

The Mystery Bags Game

Intent

This activity introduces a model for thinking about solving simple linear equations through symbol manipulation. Students summarize the model's procedure and begin to convert it to an algebraic method for solving simple linear equations.

Mathematics

Students learn a **mathematical model** for thinking about solving linear equations. The features of the pan-balance model correspond to the mathematical steps of solving a linear equation.

- The weight of a mystery bag corresponds to the unknown.

- The number of mystery bags corresponds to the coefficient of the unknown.

- The weights correspond to constants.

- Removing bags and weights corresponds to subtracting from both sides of the equation or to dividing both sides of the equation by the same amount.

Progression

Students read the introduction and do one or two examples as a class before working in small groups. The teacher introduces an algebraic representation of the game, with some student practice. Next, students describe their general strategy in words and convert this method into a symbolic representation of the steps necessary to solve the equations.

In *More Mystery Bags*, students will practice these techniques, continuing to emphasize the model to justify the steps of solving linear equations. They will also return to equations generated and solved graphically in recent activities in order to solve them symbolically.

Approximate Time

30 minutes

Classroom Organization

Groups, followed by whole-class discussion

Doing the Activity

Most students have likely already been shown how to solve simple linear equations, though many will have, at best, a procedural sense of these techniques. The introduction of the pan-balance model for solving equations is not so much to provide the procedures but to support students' own reasoning in the symbol manipulation of solving equations.

Take a few minutes to introduce the pan-balance model. You might begin by having students read the activity through to the end of the section "The Game" and then asking one or two students to describe, in their own words, what's going on. Make sure students understand these key features:

- Each mystery bag weighs the same amount.

- The pans on the two sides balance exactly.

Have someone explain how to determine the weight of the bags in Question 1: divide the weights on one side (51 ounces) by 3 (because there are three mystery bags on the other side) to get the weight of each bag (17 ounces).

Read through the remaining instructions, emphasizing that students are not only to solve the puzzles but also to explain how they know they are correct. Encourage them to use diagrams in their explanations.

Have students work on Question 2 privately. Ask for volunteers to state their solutions and explain how they know they are correct. When someone mentions removing 42 ounces of weights from each side, ask for him or her to demonstrate (or draw) this on the board. Ask the class for agreement on whether this move is allowed.

With the diagram and the statement "remove 42 ounces from each side" on the board, turn the work over to groups.

Groups will likely have a variety of solution methods. Encourage them to use diagrams to help visualize the situations, and remind them to show how they know that their conclusions are true.

Discussing and Debriefing the Activity

When all groups have completed Question 5, bring the class together and have various groups present solutions. The problems are fairly straightforward, but it's worth reviewing each in detail in order to establish the intuitive principles for working with equations. Keeping in mind that students' work with equations will gradually become more algebraic and abstract, encourage students to describe what they are doing directly in terms of the physical situation rather than using algebraic language.

For instance, for Question 3, students might begin by removing 10 ounces of weights from each side, leaving 8 mystery bags on one side and 80 ounces of weights on the other side. They can then divide by 8 to get the weight of each bag.

Question 4 is the first balance with mystery bags on both sides. Take care to restate the presenter's language about removing 3 bags from each side, leaving just 29 ounces of weights on one side and 1 bag on the other side.

Question 8 asks students to share their procedures for figuring the weight of the mystery bags. Record a summary of their procedures on the board. It is likely to look something like this:

- Remove the same number of mystery bags from each side.

- Remove the same number of ounces from each side.

- When one side contains only mystery bags and the other side contains only ounces of weights, divide the number of ounces by the number of mystery bags.

As time allows, ask for comments on Question 6 (for which no mystery bag weight could make the scale balance) and Question 7 (in which the mystery bags could weigh anything and the scales would still balance).

Next, ask the class to write a symbolic representation of the diagram for "The Game," perhaps using M for the weight of a mystery bag. Students will probably come up with either $M + M + M = 12$ or $3M = 12$.

Then ask for a volunteer to show an algebraic representation for Question 3.

Next, have groups come up with algebraic representations for Questions 1 through 7 without solving them yet. When they are done, assign one question to each group and have a representative present the group's equation to the class.

Discuss how to record the arithmetic operations of the steps in the general strategy developed earlier. Ask, **How could you record algebraically the arithmetic steps you followed to solve for the weight of a mystery bag?** For example, the first step in Question 5 could be to remove 4 mystery bags from each side, which can be recorded as follows.

$$
\begin{array}{rl}
11M + 65 & = 4M + 100 \\
-4M & \quad -4M \\
\hline
7M + 65 & = 100
\end{array}
$$

With student input, develop the remaining steps.

$$
\begin{array}{rl}
7M + 65 & = 100 \\
-65 & -65 \\
\hline
7M & = 35
\end{array}
$$

$$
M = \frac{35}{7}
$$

$$
M = 5
$$

If students instead suggest removing 65 ounces of weights from each side first, demonstrate the algebraic solution in this order. If both sequences are proposed, ask whether it matters which step is done first.

Some students may suggest that $11M - 4M$ is equal to 7, believing they can "cancel out" the M's. If this common mistake arises, it offers a great opportunity to return to the model to clarify the meaning of each algebraic symbol. Doing so should emphasize that removing 4 mystery bags from 11 mystery bags leaves 7 mystery bags, which is represented by $7M$, not by 7 ounces of weights.

Key Questions

How can you represent Question 3 algebraically?

How could you record algebraically the arithmetic steps you followed to solve for the weight of a mystery bag?

More Mystery Bags

Intent

This activity offers students more experience with the pan-balance model and encourages them to extend this reasoning toward a more general strategy for solving equations.

Mathematics

Students continue work with simple linear equations using the pan-balance method. The method is then generalized into more formal algebraic principles for solving equations. The concept of equivalent equations is also developed.

Progression

Students work individually with the pan-balance model before returning to equations identified in previous activities to solve them symbolically and confirm their solutions. Then, in a class discussion, students are introduced to a more formal consideration of equivalent equations as the broad idea that governs techniques for solving equations.

Approximate Time

5 minutes for introduction

20 minutes for activity (at home or in class)

20 minutes for discussion

Classroom Organization

Individuals, then groups, followed by whole-class discussion

Materials

Posted equations from *Water Conservation*, *The Big Buy*, and *Getting the Gold*

Doing the Activity

Tell students that they will be working through more mystery bag puzzles on their own, and remind them of the importance of explaining their solutions.

Discussing and Debriefing the Activity

Begin class with a few minutes in small groups, giving students an opportunity to share ideas and question one another about the problems. Then initiate the class discussion by going over some of Questions 1 through 6. Encourage both drawings of scales and algebraic representations of the solution steps.

Ask for a volunteer to begin the discussion of Question 7. The main point in this example is reasoning through combining the 5M and the 2M. If the first presenter solves the problem by saying, "Remove 2 mystery bags from each side, then remove 5 mystery bags from each side," ask for another volunteer who solved the problem a different way. Students should come away understanding that 5M + 2M can be rewritten as 7M because 5 mystery bags on the same side as 2 mystery bags is equivalent to 7 mystery bags on that side.

Have a few students share the equations they made up for Question 8. Then move on to Questions 9 through 11 to extend beyond the physical model to more abstract reasoning about the nature of equations. To ground this extension in students' experience, have them look back at the equations they created in their discussions of recent activities.

Question 9 is the equation from *The Big Buy*. As needed, review the problem setting and how students came up with the equation $10 + 5x = 40 + 3x$ for determining how many hours of work would yield the same amount of money earned by both Jillian and Max. Once students find the solution by using algebra, have them check it in the original equation and confirm that the answer is the same as the one they found earlier.

Review the contexts for the equations in Questions 10 and 11:

- In *Water Conservation*, students developed the equation $50 - 3x = 100 - 8x$ to determine the number of days after which the Stevens family and the Muster family would have the same amount of water remaining.

- In *Getting the Gold*, students developed the equation $15x - 60 = 3x - 420$ to determine the number of days after which the panning method and the trough method would yield the same profit.

Ask students how to use the pan-balance method, or a more abstract approach, to solve these equations, turning the question over to the groups to complete if necessary. Be sure students not only check solutions with one another but also the diagrams they drew and the symbolic steps they used.

These equations involve subtraction; they don't fit the pan-balance model as neatly as other equations have. But some students will likely see that they can either adapt that method or generalize what they have done abstractly. Once the class has solved these equations, have students verify that their solutions work and check that these are the same solutions they found earlier using different methods.

You can use this discussion as a lead-in to more formally talking about the concept of **equivalent equations**, the theoretical basis for all algebraic techniques for solving equations. The terminology described here provides a framework for discussing these techniques. You might begin by reviewing the definition of *equivalent expressions*. Ask, **What do you think the term** *equivalent* **expressions means?** Students should share ideas and possibly give examples. They may mention that substituting a value into either expression gives the same result.

Use the concept of equivalent expressions to work through the equation in Question 4. Ask for a volunteer to explain, in terms of the model, one step he or she took.

Then ask, **Why is it okay to make this change in the equation?** Point out that the goal is to solve the original equation, not some other equation. This should elicit a response like, "This equation has the same solution as the original one." Tell students that equations with the same solution are called *equivalent equations*.

Point out that some equations have more than one solution, and mention that this complicates the idea of equivalence. Because of this complication, the definition is usually formulated something like this:

Two equations are equivalent if every solution of either one is also a solution of the other.

Ask, **What are some things we can do to an equation to get a different but equivalent equation?** For each action proposed, have the class decide whether it is correct, and if so, record it on a poster. This list, like the one below, will serve to support students' subsequent work and may be added to as new ideas arise.

- Replace an expression with an equivalent expression.

- Add the same thing to both sides of an equation.

- Add the same multiple of the variable to both sides of an equation.

- Subtract the same thing from both sides of an equation.

- Subtract the same multiple of the variable from both sides of an equation.

- Multiply both sides of an equation by the same (nonzero) number.

- Divide both sides of an equation by the same (nonzero) number.

Students might also suggest some incorrect actions, or other correct actions such as squaring both sides of an equation. If so, ask for a convincing argument about why the action is correct. Challenge the class to identify counterexamples if they are uncertain about the conjecture.

Very briefly, note that for the principles involving multiplying and dividing, the number involved must be nonzero. Students already know that they can't divide by zero, so the restriction on division shouldn't cause confusion.

Key Questions

What does that step mean in terms of the situation?

How can you use the pan-balance method or a similar technique to solve these equations algebraically?

What do you think the term *equivalent expressions* means?

Why is it okay to make this change in the equation?

What did you do with the mystery-bag problems to get equivalent equations?

What are some things we can do to an equation to get a different but equivalent equation?

Scrambling Equations

Intent

This activity will strengthen students' understanding of equivalent equations and some of the symbol-manipulation processes that maintain this equivalence.

Mathematics

Students create linear equations by "scrambling" an equation of the form $x = c$, using the rules developed in class for maintaining equivalent equations. Students take a simple equation, such as $x = 5$, and write a series of more complex equations in which each is equivalent to the preceding one. The focus thus shifts from simplifying a complex equation to "complexifying" a simple one, emphasizing the "doing" and "undoing" aspects of algebraic thinking.

Progression

Students work in groups to turn simple equations into more complex, equivalent equations and then try to retrace one another's steps by "unscrambling" their equations. This work is followed up in *More Scrambled Equations and Mystery Bags*.

Approximate Time

25 minutes

Classroom Organization

Groups

Materials

Index cards (optional)

Doing the Activity

Ask a volunteer to read the introductory paragraph. Have students read through the series of four equivalent equations, and then ask another volunteer to read the second paragraph. Instruct students to take a couple of minutes in their groups to identify what was done to each equation to get the one below it. You might follow up by re-creating the steps on the board, with student input as to what was done to each equation to get the next.

Tell groups that they are responsible for reading the remaining instructions. (The "precise rules" presented in this activity are exactly those that arose in the class discussion of *More Mystery Bags*.)

Index cards may be a little more convenient for students to record their scrambled equations than full sheets of paper. Emphasize that the cards should be passed "scrambled" side up. The author's name should also be on the card. The receiver of the card is to figure what simple equation began the process.

As you monitor the "unscrambling" process, if some students seem stumped, or if some have found errors (there will likely be many), encourage them to share their thinking with the authors.

Discussing and Debriefing the Activity

If time allows, it is worth reviewing an example or two. You might have a few students demonstrate what they did to "unscramble" specific equations.

Bring out two key points: (1) Each step they followed was the reverse of a step that the author used in creating the equation. (2) They did these "backward" steps in the reverse order from the order of the original steps.

More Scrambled Equations and Mystery Bags

Intent

This activity further develops students' abilities to simplify and solve equations. At the conclusion of this activity, students will have two related ways to think about solving equations. The pan-balance model provides a more concrete metaphor for manipulating symbols while the set of rules developed to maintain equivalent equations is useful in more varied situations.

Mathematics

The algebraic thinking developed during this unit emphasizes reasoning, multiple representations, and the connections among these representations. By this point in the unit, students will have solved a variety of equations, using both the pan-balance model and some general principles for symbol manipulation that maintain equivalent equations. These symbol-manipulation methods should be thought of as complementing other techniques students have for solving equations, including using a graph, a table, estimation, and other reasoning processes.

Progression

After students work on their own to use the rules of equivalent equations and the pan-balance model, they come together to share results and ask questions of one another. The teacher emphasizes the "doing" and "undoing" nature of these problems, some of which may remain unsolved by some students at this time.

Approximate Time

5 minutes for introduction

30 minutes for activity (at home or in class)

20 minutes for discussion

Classroom Organization

Individuals, followed by whole-class discussion

Doing the Activity

Tell students that they will be unscrambling a few more equations in this activity, following the rules for keeping equations equivalent. They will also try to solve a couple of mystery-bag problems, including one that may be especially challenging. Students should record, in some way, what they do at each step to unscramble the equation or solve the mystery-bag puzzle.

You may want to mention that Questions 2 through 4 were not necessarily scrambled in exactly three steps, as was student work in *Scrambling Equations*.

Don't expect all or even most students to be able to complete Question 3, 4, or 7 on their own at this point. The techniques they have been developing, especially as applied to more difficult problems, will be strengthened and refined in future work.

Discussing and Debriefing the Activity

The follow-up to this activity could begin with groups sharing results and asking questions of one another.

Then ask a few volunteers to describe what they did to unscramble one of Questions 2 through 4. It is not necessary that all students are confident with the procedure for Question 4, but they may be curious to see it solved if they haven't figured it out yet.

During the discussion, reemphasize that each step students took is essentially the reverse of one of the steps that was used to create the equation and that they did these "backward" steps in the reverse order from the order of the original steps.

Family Comparisons by Algebra

Intent

In this activity, students reexamine the questions raised in *Following Families on the Trail* and apply their new understanding of equations and equation-solving techniques to those questions. This activity concludes the part of the unit devoted to solving equations algebraically and emphasizing the connection to other types of solution methods. It will help you evaluate students' ability to represent situations using linear equations as well as their facility with solving them.

Mathematics

Students create equations that were originally derived in *Following Families on the Trail* and solve them using the algebraic methods they have been developing. Each situation had previously been modeled by two linear functions and solved by determining the coordinates of the point of intersection of their graphs.

Progression

Students work on the activity in small groups. Subsequent class discussion reemphasizes important ideas from the past several activities and helps students to confirm their understanding of methods for solving simple linear equations.

Approximate Time

25 minutes

Classroom Organization

Groups, followed by whole-class discussion

Materials

Students' work from *Following Families on the Trail*

Doing the Activity

Tell students that in this activity they will return once more to a previous activity and solve a question already asked, this time using the algebraic techniques they have recently studied.

Ask students to work in their groups to complete this activity. Give some groups transparencies to prepare to present either Question 1 or 2.

If groups have difficulty setting up an equation, help them focus on two expressions that must be set equal. **What two things are you trying to set equal?** They could write this in words; for example,

Buck's distance from Green River = Wood's distance from Green River

If necessary, you might ask a follow-up question like, **What arithmetic do you have to do to figure the Buck family's distance from Green River after 5 days?** Then explain that if students generalize that procedure for X days, they will have an expression for the left side of their equation.

Discussing and Debriefing the Activity

Although Question 1 may seem straightforward, a presentation will provide a good opportunity to summarize ideas about both creating and solving equations. After the presentation, ask, **In what ways does your solution to this algebraic equation relate to the graphs you built in *Following Families on the Trail*?**

Encourage students to turn back in their notes to these graphs, either individually or in their groups, and look for connections. After a couple of minutes, ask for a volunteer to address your question.

On the surface, Question 2 appears quite similar, but the fact that the two families consume coffee at the same rate may create confusion. Students are likely to arrive at the equation $100 - 5x = 70 - 5x$, but become perplexed when they try to solve it.

This is a good time to look back at the graphical representation of the situation. Students likely saw that the graphs for the two families were parallel lines, meaning the families would never have the same amount of coffee. Similarly, there is no value for x that gives $100 - 5x$ and $70 - 5x$ the same value.

Help students see that using the usual algebraic methods yields $100 = 70$, and explain that this essentially means there is no solution to the original equation.

Key Questions

What two things are you trying to set equal?

What arithmetic do you have to do to figure the Buck family's distance from Green River after 5 days?

In what ways does your solution to this algebraic equation relate to the graphs you built in *Following Families on the Trail*?

Starting Over in California

Intent

This activity provides some challenging problems in a linear context for students to solve in any manner that makes sense to them. The work will refresh students' ideas about much of the unit's mathematics and help prepare them for the portfolio and unit assessments.

Mathematics

The first two questions provide information about a linear situation. Students will need to apply the tools developed throughout the unit to solve them, including some or all of these important ideas: rate of change, starting value, In-Out tables, graphs, rules, and algebraic solution methods. Students are then asked to develop a situation and problem of their own.

Progression

Students will work individually to analyze and make predictions about linear situations and to write their own problems. When they come together to share their work, they will have the opportunity to see a variety of approaches to each problem and to share their own—and solve another's—problem.

Approximate Time

30 minutes for activity (at home or in class)

20 minutes for discussion

Classroom Organization

Individuals, followed by whole-class discussion

Doing the Activity

The activity requires little introduction. Encourage students to be sure they can solve the problem they create for Question 3, as they will be sharing it with another student.

Discussing and Debriefing the Activity

You might have students begin by comparing answers and methods for Questions 1 and 2 in their groups. Then ask for volunteers to show how they solved these somewhat challenging problems; it is likely that students will have used a variety of methods.

Conclude the activity by having students pair up and exchange their problems for Question 3. Provide time for them to solve and check solutions with one another.

Beginning Portfolios

Intent

Students begin work on their unit portfolios. In the process, they begin to review the unit, reflect on specific aspects of what they have learned, and study for the unit assessments.

Mathematics

Students look back on their work with graphs to identify two ways in which they were used: to describe or model a problem situation and to make a decision about a problem situation.

Progression

Students work on this activity individually.

Approximate Time

20 minutes

Classroom Organization

Individuals

Materials

Students' work from the entire unit

Doing the Activity

Prior to this beginning work on the unit portfolios, remind students to bring all their work from the unit to class.

Discussing and Debriefing the Activity

Encourage students to share some of their thoughts about the period of the gold rush with the class, perhaps asking them to read portions of what they wrote.

You might also ask volunteers to identify the names of the activities, and possibly question numbers, they selected for each graph.

The Overland Trail Portfolio

Intent

Students review and document their mathematical activity and learning during the course of the unit. Their product is an opportunity for assessing what they have learned and what they believe is important in their learning (see "Portfolios" in *A Guide to IMP* [link to the guide]).

Mathematics

Students review their work on all the mathematical topics of the unit, with an emphasis on constraints and decision making; algorithms, variables, and expressions; graphs; linear functions and equations; and multiple representations of linear relations (situations, tables, graphs, and equations).

Progression

Students are introduced to the particular expectations for *The Overland Trail* portfolio. They review their materials and begin to compile their portfolios as they write their cover letters.

Approximate Time

10 minutes for introduction

50 minutes for activity (at home or in class)

Classroom Organization

Whole-class introduction, then individuals

Doing the Activity

Although this is primarily an activity to be done on one's own, some time in class can be quite productive as students work together to identify some of their work they would like to include. The portfolio-assembly process can be a chance to share pride in the work done during the unit.

This will be students' third portfolio. By now students understand the value you place on portfolios in terms of their learning (and possibly their course grade). However, it is likely they still don't have a good sense of what might make a good portfolio and how to go about constructing one.

Adapt the suggestions below to the in-class and out-of-class time you have allotted for students' work on their portfolios. One goal for this portfolio is to help students make significant improvements in the quality of the portfolio as a whole and of each of the three parts. You might spread some of this work out over a few days.

Ask students what they have found valuable in completing other portfolios. Look for replies like "It helped me study for the test," "It helped me realize what I have learned," and "I like to look back over the work I've done well." Remind students that a portfolio is meant to give them an opportunity to demonstrate both the variety of things they are able to do mathematically and the quality with which they can do these things.

Tell students that with those purposes in mind, you want to review the components of an IMP portfolio—writing the cover letter, choosing papers, and writing the personal growth—one more time, and have them read over the instructions in the student book carefully.

Explain that it makes the most sense to begin with the second step, choosing papers. Remind them to make notes of why they decided to select the activities they did, as these thoughts need to appear in their cover letters.

After some time has been spent on that task, turn students' attention to the task of writing about their personal growth. In *The Overland Trail*, this section will be, primarily, about reflecting on graphing. Invite students to expand on this requirement if they feel strongly about any other personal growth they experienced during the unit.

Finally, students should write a cover letter that describes the main mathematical ideas of the unit and how they were developed, as well as introduces and explains why they are presenting the papers they have selected.

You might also review the main mathematical ideas of the unit. Some teachers post these to begin a unit or add to a poster as the unit progresses. Ask students what the main ideas were and help them to summarize them into about five "big ideas." Students can base their own writing on this organization.

Discussing and Debriefing the Activity

More than in previous units, this could be a good opportunity for students to share their portfolios. They might show them to their group members, comparing ideas about the mathematical topics written about in the cover letters and the activities they have selected.

If you create a simple rubric for the portfolio's expectations, students could check their group mates' portfolios to ensure that each person completed each component. You might elect to do this on the day the portfolio is due and then allow those students who desire it one more day to complete any portion of the portfolio that a group member noticed was incomplete.

Wagon Train Sketches and Situations

1.

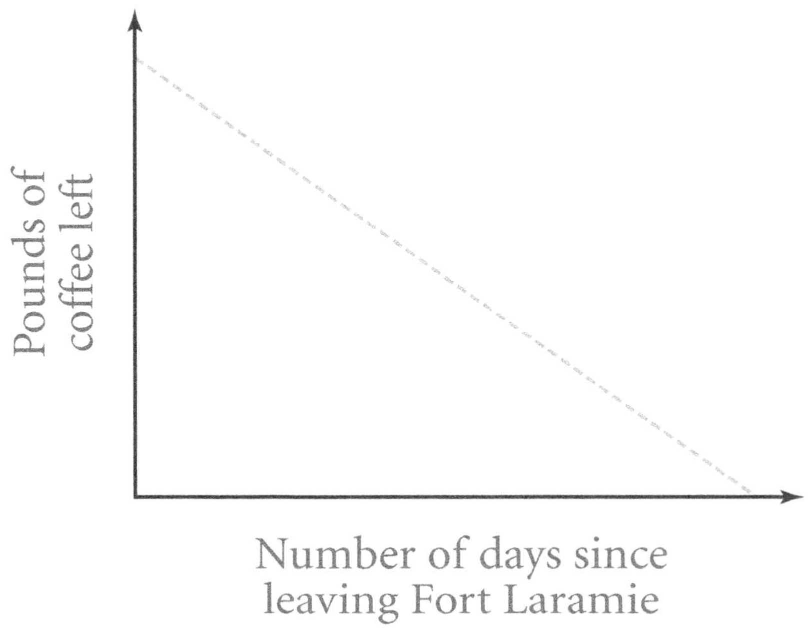
Pounds of coffee left
Number of days since leaving Fort Laramie

2.

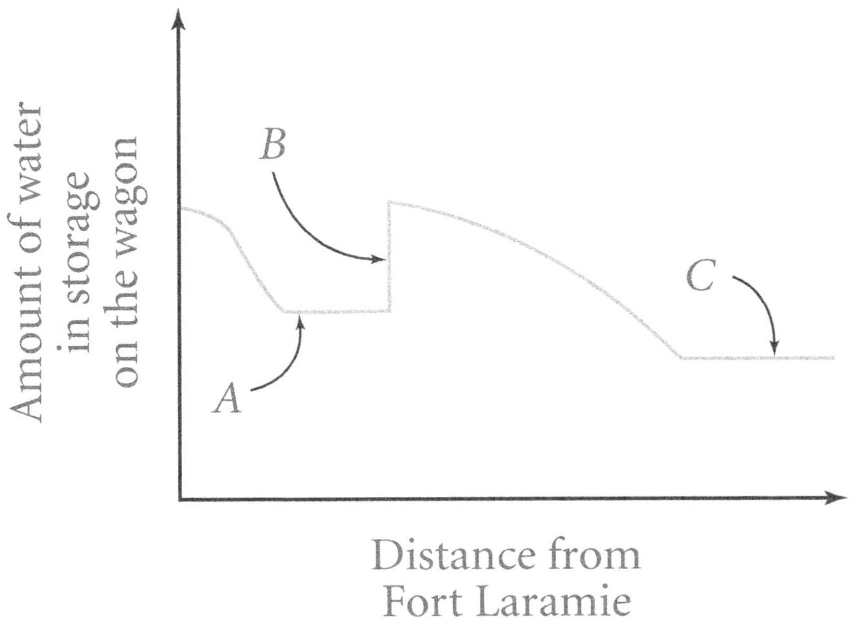
Amount of water in storage on the wagon
B
C
A
Distance from Fort Laramie

Wagon Train Sketches and Situations (continued)

3.

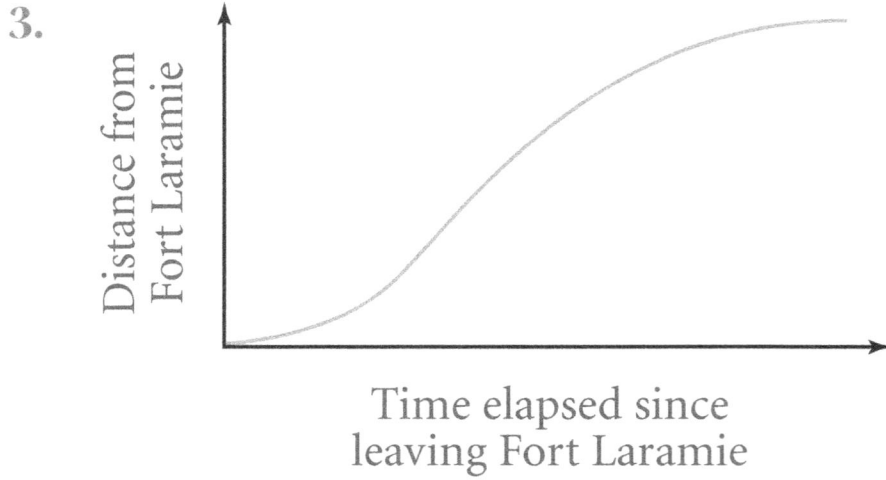

Distance from Fort Laramie

Time elapsed since leaving Fort Laramie

4.

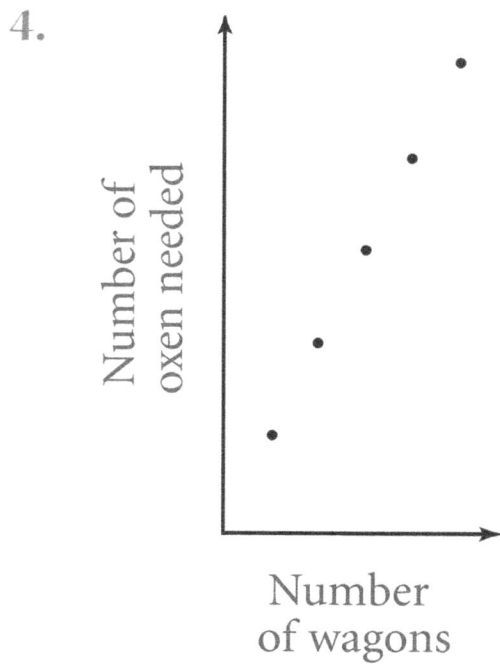

Number of oxen needed

Number of wagons

Graph Sketches

1.

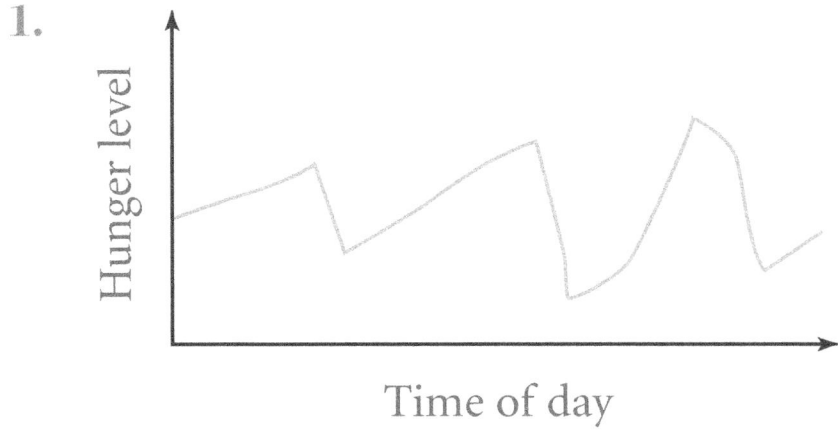

Hunger level

Time of day

2.

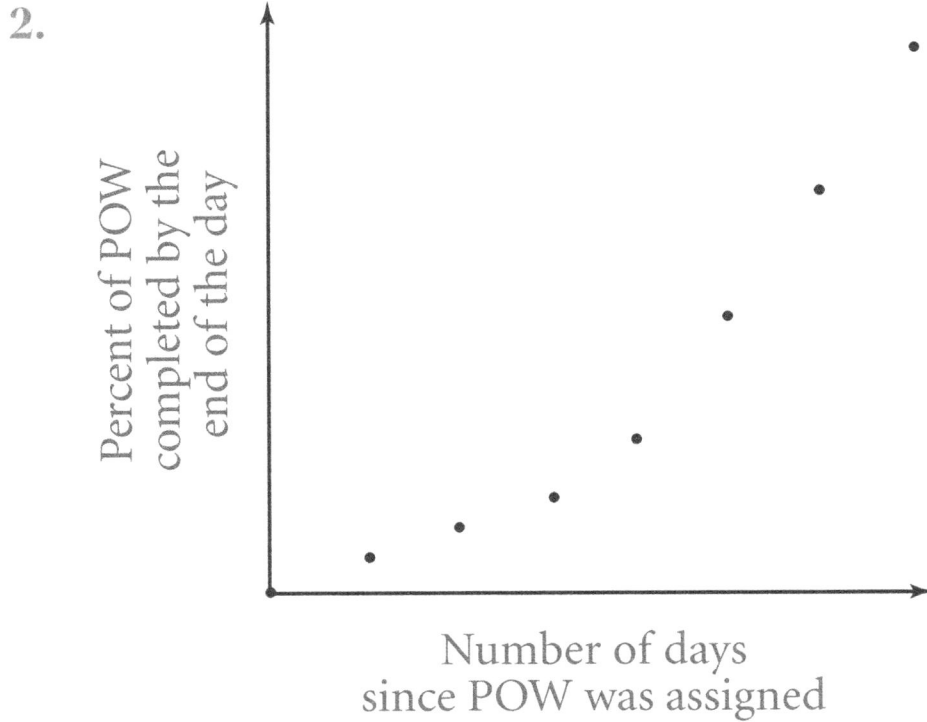

Percent of POW completed by the end of the day

Number of days
since POW was assigned

Graph Sketches (continued)

3.

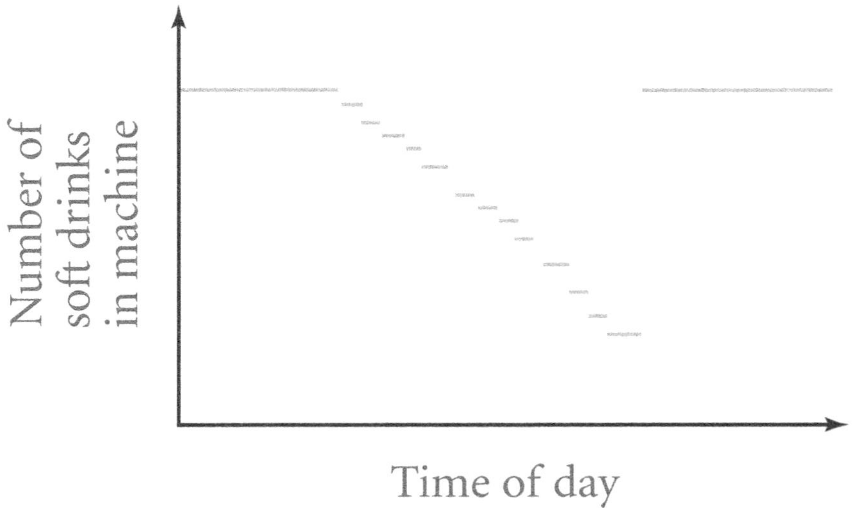
Number of
soft drinks
in machine
Time of day

4.

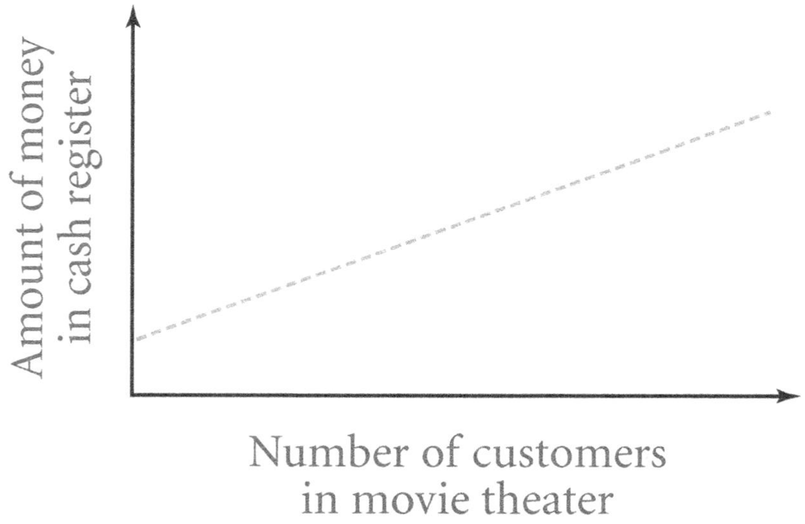
Amount of money
in cash register
Number of customers
in movie theater

In Need of Numbers

1.

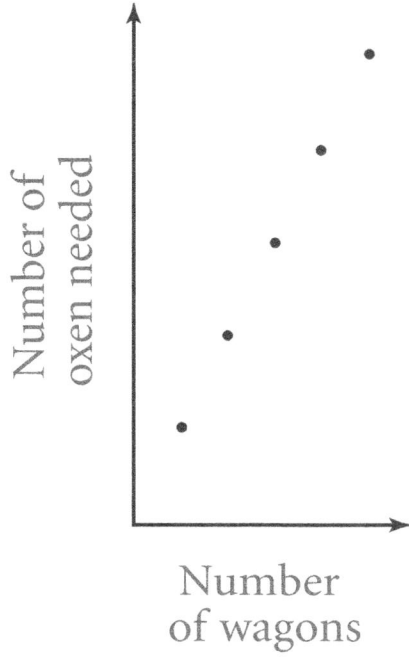

Number of oxen needed

Number of wagons

2.

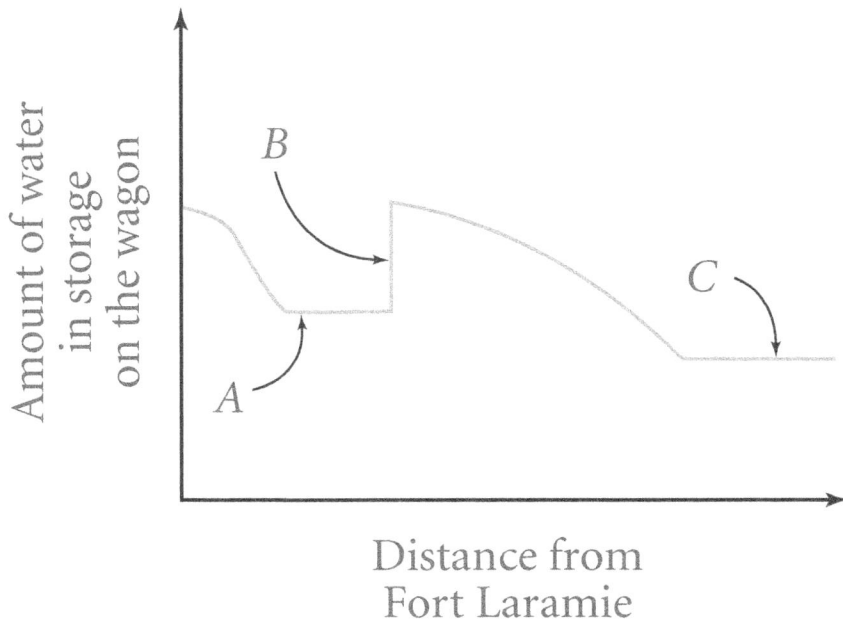

Amount of water in storage on the wagon

B

A

C

Distance from Fort Laramie

In Need of Numbers (continued)

3.

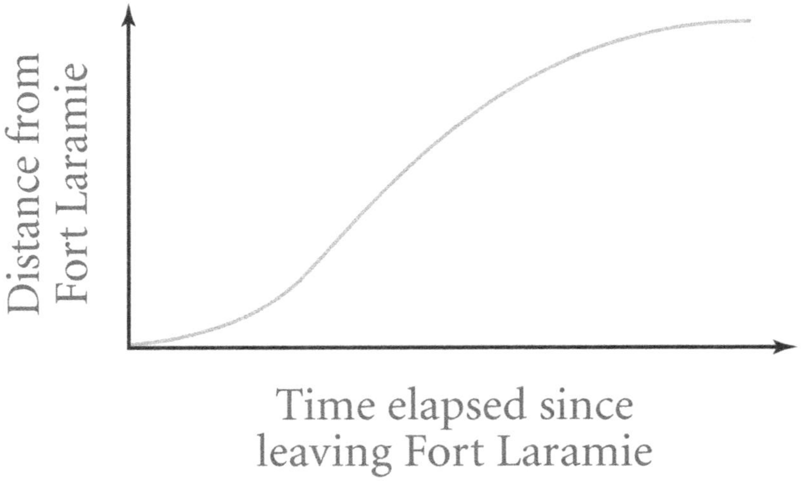
Distance from
Fort Laramie
Time elapsed since
leaving Fort Laramie

4.

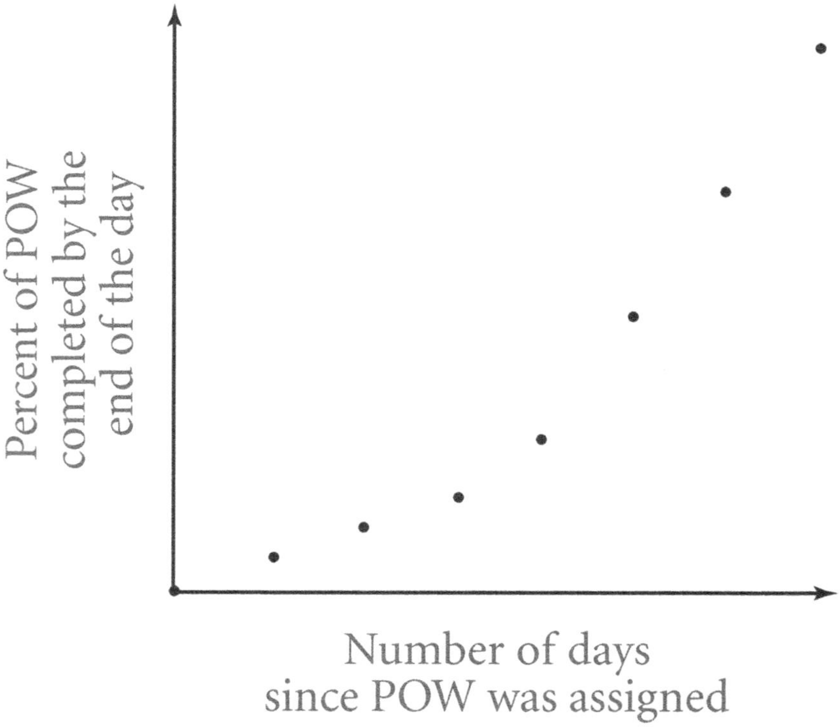
Percent of POW
completed by the
end of the day
Number of days
since POW was assigned

In Need of Numbers (continued)

5.

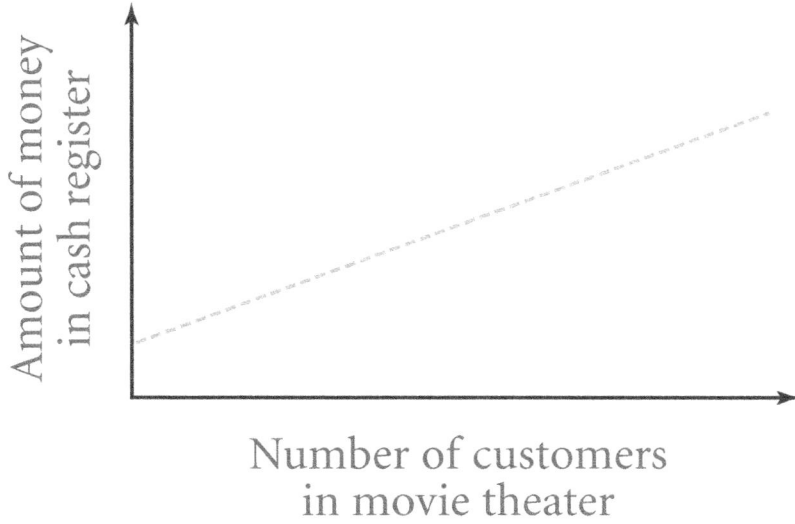

Amount of money in cash register

Number of customers in movie theater

The Issues Involved

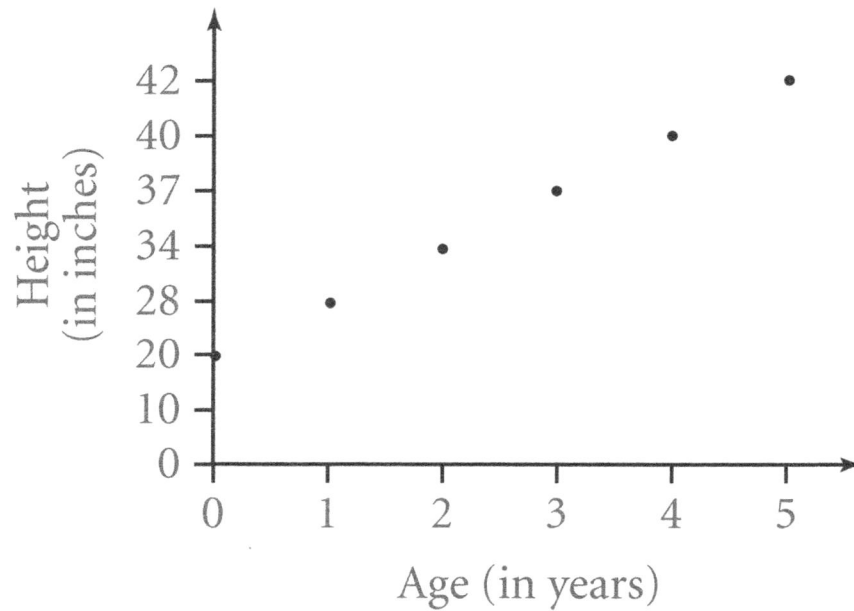

Out Numbered

1.

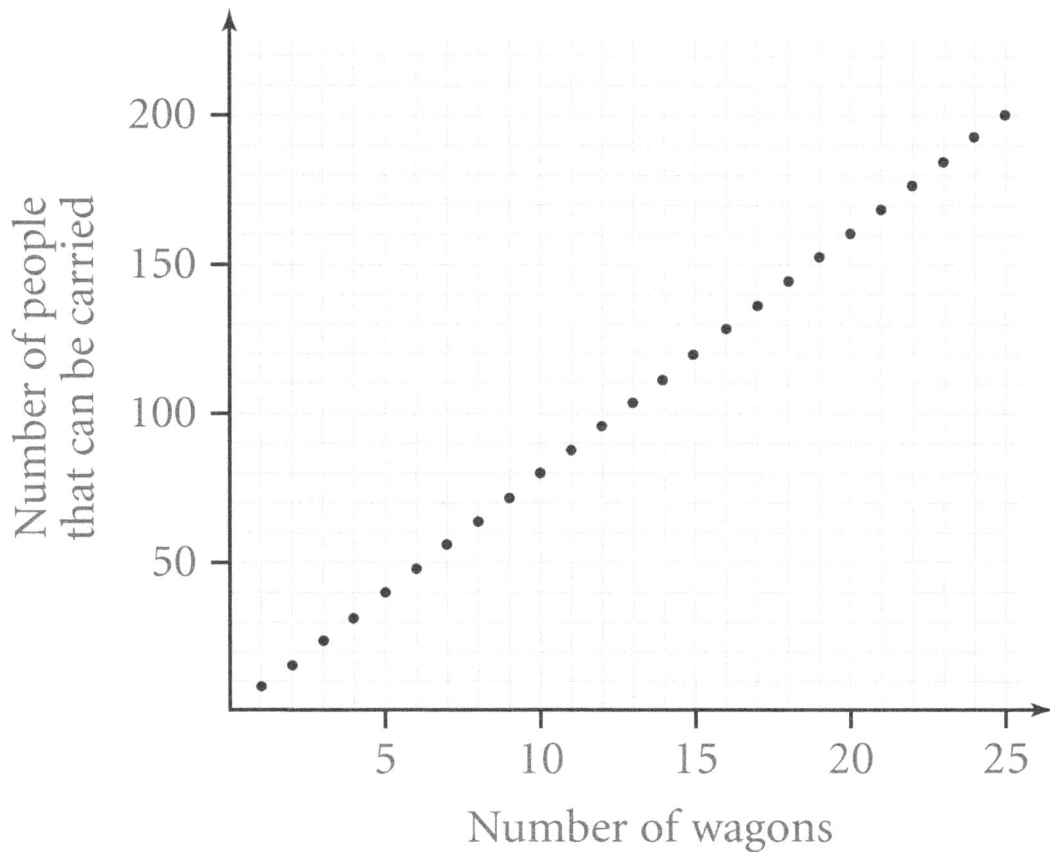

Number of people that can be carried (y-axis) vs. Number of wagons (x-axis)

2.

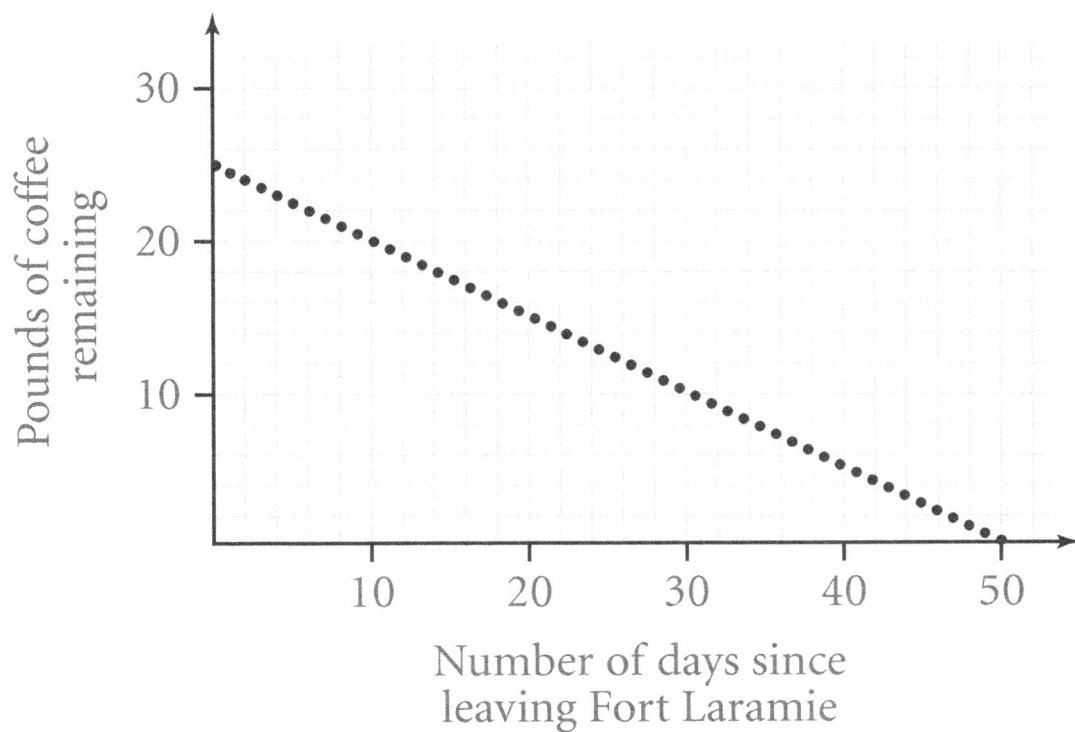

Pounds of coffee remaining (y-axis) vs. Number of days since leaving Fort Laramie (x-axis)

Out Numbered (continued)

3.

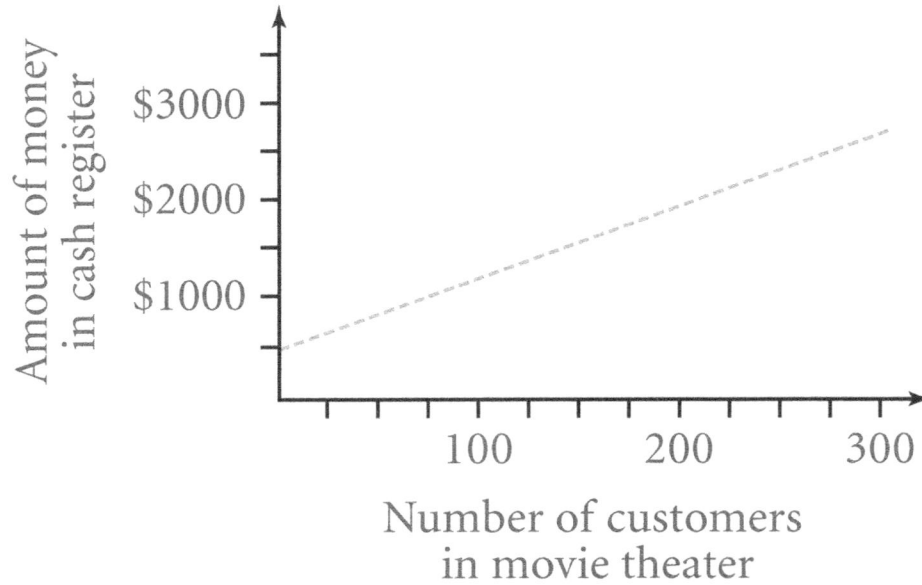

Broken Promises

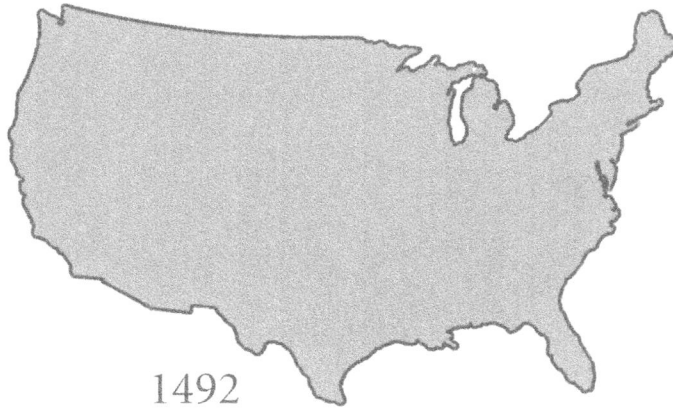

1492

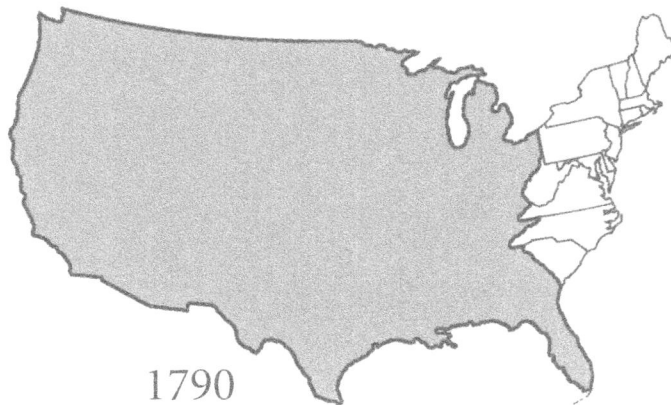

1790

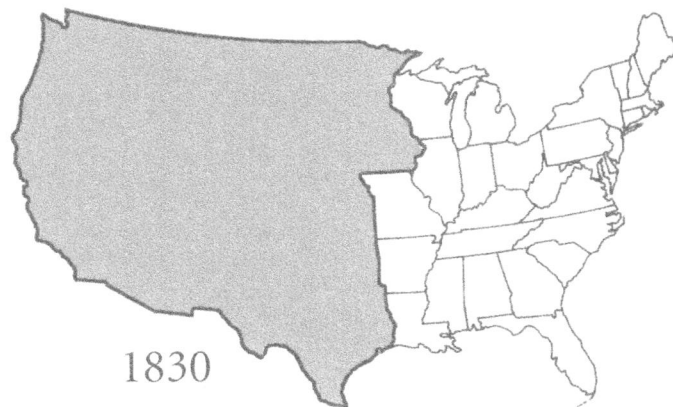

1830

Broken Promises (continued)

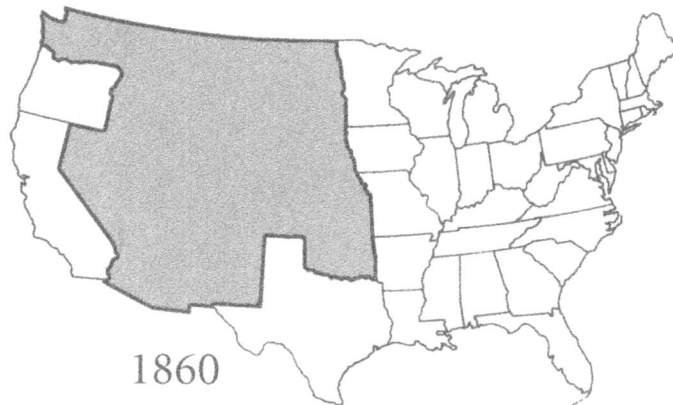

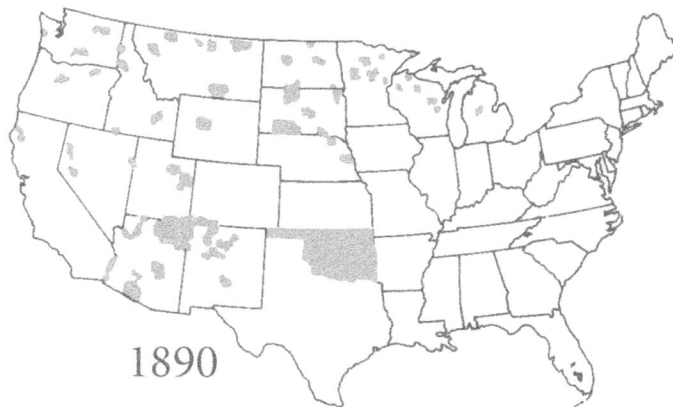

1860

1890

1-Centimeter Graph Paper

$\frac{1}{4}$-Inch Graph Paper

1-Inch Graph Paper

These questions are about a modern-day move from Kansas City, Missouri, to San Francisco, California.

Unlike the settlers of 1850, your family will travel on superhighways with a large rental truck and a van. The distance between the two cities is 1800 miles.

In answering the questions, be sure to show your reasoning, not just give an answer. Assume that your average speed is 50 miles per hour.

1. Suppose you estimate that you can drive an average of six hours on a typical day of the trip. At this rate, how many days will the trip last?

2. Construct an In-Out table for this situation. The *In* column contains the number of hours of driving per day. The *Out* column shows how many days the trip will last.

3. Plot the points from your table to show the number of days in the trip as a function of how many hours you are willing to drive per day.

4. Create an algebraic equation that shows how to compute the number of days required for a trip like this (not necessarily 1800 miles—you should represent the distance by a variable). This expression should not contain any numbers, and you should clearly define the symbols in it.

As in the In-Class Assessment, these questions below have to do with a modern-day move.

Your family is traveling on superhighways with a large rental truck and a van.

1. Fuel Economies

Each time one of the vehicles stops for gas, you make a note of how much gas the vehicle has used and how many miles it has traveled.

a. Using the data in the tables below, draw two graphs *on the same set of axes* comparing total miles traveled to total gallons of gas used for each vehicle.

Van		Truck	
Total gallons of gas used so far	Total miles traveled so far	Total gallons of gas used so far	Total miles traveled so far
18	400	36	300
28	700	65	600
42	990	100	850
52	1250	138	1100

b. Reading from the graph, estimate how many gallons of gas the truck had guzzled by the time you traveled 800 miles. Do the same for the van.

c. Estimate how many miles each vehicle can travel on a gallon of gas.

2. Van Repairs

One morning, you discover that the van won't start. The mechanic says that the fuel pump needs replacement and that the work can be done in about two hours.

The family decides that you and one parent will head out in the truck toward that night's stop, 400 miles down the highway. The rest of the family will remain behind and either catch up with you during the day or meet you tonight at the motel where you have reservations.

a. Suppose the truck travels 45 miles per hour, the van goes 55 miles per hour (once it is fixed), and the repair takes two hours. For simplicity, assume that neither vehicle makes any stops along the way.

      i. Which vehicle will reach the motel first?

      ii. How far from the motel will the other vehicle be when the first vehicle arrives?

b. Of course, the repair might not take the full two hours, or it might take longer. So now consider the case in which the repair actually takes four hours. In this case, the truck definitely arrives first.

    How far from the motel will the van be when the truck arrives?

c. Now generalize from parts a and b. Suppose that h represents the number of hours the van is delayed. Assume that h is a number large enough to allow the truck to arrive first.

    Develop a formula or an equation in terms of h that tells how far the van will be from the motel when the truck arrives.

The Overland Trail Guide for the TI-83/84 Family of Calculators

This guide gives suggestions for selected activities of the Year 1 unit *The Overland Trail*. The notes that you download contain specific calculator instructions that you might copy for your students. NOTE: If your students have the TI-Nspire handheld, they can attach the TI-84 Plus Keypad (from Texas Instruments) and use the calculator notes for the TI-83/84.

Students study graphs of functions in this unit. Using technology, they can graph many functions quickly and compare them easily. You'll find that calculator use ties in naturally with the existing curriculum and is more than simply an interesting extension. Technology allows students to move among different representations of a function: as a table of values, as a graph, and as an algebraic equation.

When students find lines to fit their data sets, they can experiment in the calculator's Y= menu to find best-fit lines of their own design. Near the end of the unit when students look for a common solution to a pair of linear equations, they can graph the equations and look for the point of intersection.

"Graphing Basics" introduces the five graphing keys positioned directly below the calculator screen. These instructions are general enough that they make good reference material throughout the unit.

The Search for Dry Trails: During the review of this activity, you will discuss the mean and median of a data set. Students might wonder how to calculate the mean and median of a data set using the calculator. The instructions for "Calculating the Mean and Median of a Data Set" are in the notes.

The Issues Involved: Your discussion of this activity will deal with misleading scales on graphs. Because a graphing calculator can only scale each axis uniformly, you cannot use a graphing calculator to replicate the distortion of the data in the graph from *The Issues Involved*. However, because it is easy to adjust the window to investigate different scales, the graphing calculator is a good tool for discussing misleading scales. A misleading window on the calculator's small screen is a common reason students misinterpret graphs. If you would like more details on how to use the different graphing keys, look at "Graphing Basics."

From Rules to Graphs: In the rest of the activities in the textbook section "The Graph Tells a Story" and the first six activities in the textbook section "Traveling at a Constant Rate," students study these topics: graphing equations of functions, plotting data points, creating data tables using a function rule, and finding lines of best fit for a data set. Instead of waiting until *Graphing Calculator In-Outs*, you might choose to integrate calculator use with the new material that leads up to the calculator activity. Or, because students can also be overwhelmed by learning new material and new keystrokes at the same time, you might prefer to wait until *Graphing Calculator In-Outs* to formally introduce calculator use with these topics.

Graphing Calculator In-Outs: The main purpose of this activity is to familiarize students with the graphing capabilities of their calculator. "Graphing Basics" contains general instructions about using the five graphing keys. You may want your students to have "Graphing Basics" on hand throughout the rest of the unit. The section that follows, "Help with Graphing Calculator In-Outs," helps students through Question 1 in the activity *Graphing Calculator In-Outs.* Students can then work through the other questions using this discussion as a reference.

"Sublette's Cutoff" Revisited: For this activity, students might use the notes on "Finding a Line of Best Fit and Using It for Prediction." They show how to find a best-fit line through experimentation. The instructions do not include steps for using the calculator's regression features to find a best-fit line. This omission is intentional. Students learn so much by experimenting on their own to find best-fit lines that many teachers don't want them using the calculator's statistical capabilities right away.

Fair Share on Chores: In this activity, students use graphing calculators in Questions 5 and 6. You could have available the notes on "Graphing Basics" for students to use as a reference. This activity is also a good opportunity to let students explore using a table of values, as described at the end of "Help with Graphing Calculator In-Outs." This is a relatively simple extension and should not take a lot of time.

More Fair Share for Hired Hands: To use graphs to solve a system of equations, graph both equations and find their point of intersection. Once you have entered both functions in the Y= menu and found a reasonable window, press TRACE. You can use the up or down arrow key to move the cursor from one function to another. The TI-83/84 family of calculators indicate which function or statistical plot you are currently tracing in the upper-left corner of the screen. When the cursor is near the point of intersection of two functions, the coordinates showing at the bottom of the screen should change very little as you move up and down between the functions with the arrow keys. Zooming in on the point of intersection can improve your accuracy, and sometimes a table can help you find an exact solution. After students have had ample experience estimating intersections by tracing, you can encourage them to research the calculator's **intersect** command.

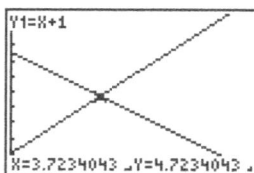

Water Conservation: In this activity, students graph each family's water supply and determine when the family will run out of water. They need to identify the places where the graphs cross the *x*-axis, or *x*-intercepts. If you graph the water supply on the calculator, you can use tracing to estimate the x-intercepts. Use the left and right arrow keys to trace along the graph and search for the spot where the y-coordinate is as close as possible to zero. Find the point as precisely as you like by zooming in. After students have had ample experience estimating x-intercepts, you might encourage them to research the calculator's zero command. You may need to explain that a "zero" is an x-value that makes the entire function equal to zero, or y = 0.

The Overland Trail Portfolio: Near the end of the activity, students describe the benefits of graphing with and without calculators. Have calculators available so students can use them to review their calculator skills as they proceed. If students want to include calculator work in their portfolios, they might want printouts of the calculator screens. To print a calculator screen, you'll need to link the calculator to a computer using the TI Connect™ cable and software. Once students transfer the graph to the computer, they can print from the computer or save image files for use in other computer applications.

Supplemental Activities

Some of *The Overland Trail*'s Supplemental Activities provide opportunities to introduce additional features of the graphing calculator.

Pick Any Answer: Challenge students to write simple programs that carry out the steps of a number trick. This can provide a good review of programming as well as a speedy way to recognize patterns in the outcomes of a trick.

Substitute, Substitute: Storing values for variables and evaluating expressions on the home screen can be used to check by-hand calculations. For exponents greater than two, you'll need to review the use of the carat symbol, ^.

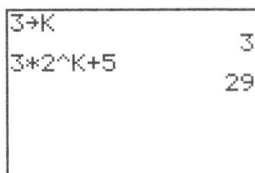

```
3→K
             3
3*2^K+5
            29
```

Integers Only: You might encourage students to research some of the calculator's integer functions, such as **iPart(**, **fPart(**, and **int(**. The latter function is the calculator's greatest integer function and can be found by pressing the **MATH** key and going to the **NUM** submenu. It can be used to evaluate expressions on the home screen or to graph a step function.

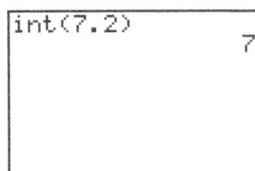

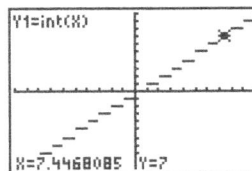

```
int(7.2)
             7
```

```
Y1=int(X)




X=7.4468085  Y=7
```

Calculating the Mean and Median of a Data Set

You can use a special feature of your calculator to find the mean and median of a set of numbers. This feature is especially helpful when you are working with a large data set. Let's use an example from *The Search for Dry Trails* and work with the data set showing the number of rainy days various travelers have observed on the Oregon Trail.

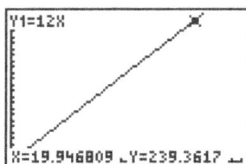

Press [STAT] [ENTER] to get to your screen of lists. You are going to enter your data in list **L1**. If this list already contains data, place your cursor on the label of the list name and press [CLEAR] [ENTER].

Now enter the six numbers in this data set. To enter each value, input the number and press [ENTER]. Then return to your home screen by pressing [2nd] [QUIT].

Now press [2nd] [LIST]. Select the **MATH** menu, highlight the **mean(** or **median(** command, and press [ENTER]. Then press [2nd] [L1]. Close your parentheses and press [ENTER]. Your calculator will find the mean or median of all the numbers in your list!

For practice, use your calculator to find the mean and median of another set of data. You do not have to clear out the data in list **L1**. Simply enter your data set in list **L2**, and then calculate **mean(L2)** or **median(L2)**.

Graphing Basics

These instructions describe the five basic graphing keys. You will find these keys immediately below the calculator screen.

Press $\boxed{Y=}$ to get the screen where you tell the calculator what functions to graph. Use the $\boxed{X,T,\theta,n}$ key for the independent variable X when you enter a function. Before you begin, press $\boxed{MODE}$ and make sure the fourth line is set to **Func**.

You remove functions from the list with the $\boxed{CLEAR}$ key.

You can also make a function inactive without removing it. To do this, place the cursor over the = sign for that function and press $\boxed{ENTER}$. The = sign will lose its highlighting. In the example just shown, the function **Y₁=12X** is active, and the function **Y₂=3X+5** is inactive. The calculator will graph only active functions. To make the function active again, put your cursor on = and press $\boxed{ENTER}$. You can also arrow further left to the line beside each function and press $\boxed{ENTER}$ to change the type of line and shading.

Press $\boxed{WINDOW}$. This is where you tell the calculator what part of the graph to show and how to scale the axes. The **Xmin** and **Xmax** values determine the left and right bounds of the graph. The **Ymin** and **Ymax** values determine the bottom and top bounds. The **Xscl** and **Yscl** values decide the frequency of the tick marks on each axis. **Xres** should be set at 1.

Press $\boxed{ZOOM}$. The ZOOM menu is another way to tell the calculator what part of the graph to show and automatically adjusts the window settings. Each feature has its own purpose. **ZStandard** is handy because it resets the window to a standard –10 to 10 setting. To use **Zoom In** or **Zoom Out**, highlight your choice and press $\boxed{ENTER}$. A blinking cursor will appear on the graph. Move the cursor to the spot you want to zoom in to or out from and press $\boxed{ENTER}$ again. **ZSquare** adjusts your window so a distance on the *y*-axis is the same length as a corresponding distance on the *x*-axis. This feature keeps your graphs from being stretched or distorted.

Continued on next page

Y1=12X

X=.5   Y=6

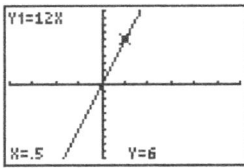

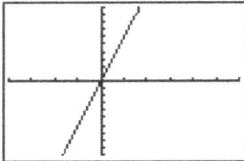

Press ⌈TRACE⌉ and experiment with the left and right arrow keys. Observe the *x*- and *y*-coordinates of the cursor at the bottom of the screen. If you have more than one function in your graph, you can use the up and down arrow keys to move the cursor between functions.

Press ⌈GRAPH⌉. This is how you tell the calculator to draw the graphs of the active functions in the ⌈Y=⌉ edit screen and stats plots. The graph will match the settings you have in the ⌈WINDOW⌉ display.

Help with Graphing Calculator In-Outs

The instructions in this section will help you use a graphing calculator to complete Question 1 from the activity *Graphing Calculator In-Outs*. Then you can work through Questions 2 and 3 using this discussion as a reference.

Look at *Graphing Calculator In-Outs*. Read the introduction and all of Question 1 before you get started. If you can already answer Questions 1b and 1c, make a note of your answers and use them to check the answers you get on the calculator.

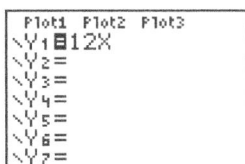

```
Plot1 Plot2 Plot3
\Y1■12X
\Y2=
\Y3=
\Y4=
\Y5=
\Y6=
\Y7=
```

This activity gives an equation expressing the number of pounds of beans in terms of the number of people in the family. First write down this equation on paper, using X for the number of people and Y for the number of pounds of beans. Then press $\boxed{Y=}$ and enter your function.

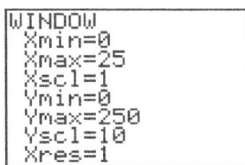

```
WINDOW
 Xmin=0
 Xmax=25
 Xscl=1
 Ymin=0
 Ymax=250
 Yscl=10
 Xres=1
```

Press $\boxed{GRAPH}$ to view your graph. Then press $\boxed{WINDOW}$ and adjust your settings so you have a good window for viewing this function. Look at Questions 1b and 1c to give you some ideas about maximum values for X and Y.

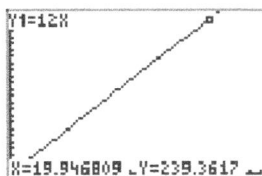

```
Y1=12X

X=19.946809  Y=239.3617
```

Now press $\boxed{TRACE}$. Experiment with the left and right arrow keys to move the cursor along the graph. Observe the coordinates of the cursor. To answer Question 1b, you want to find the point where your independent variable is 20. Trace along the function until your *x*-coordinate is as close to 20 as possible.

To get a more precise graphical answer, press $\boxed{ZOOM}$. Then choose **Zoom In**. Because your cursor is already near X = 20, that's a good spot to zoom in on. Press $\boxed{ENTER}$. You should see a magnified piece of your previous graph. Press $\boxed{TRACE}$ again and look for a value closer to X = 20. Do you see how you can continue to zoom in to locate a point more and more precisely?

Continued on next page

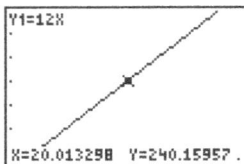

When you are given an X-value, you can find the exact Y-value by pressing [2nd] [**Trace**] and selecting **value**. Type the X-value and press [ENTER]. Your calculator gives you the Y-value. (NOTE: The value command only works from X to Y. You have to use tracing or a table to go from Y to X.)

Now you can use your calculator to answer Question 1c. You need to trace to a spot where Y = 155. You may need to **Zoom Out** first or to change the window by adjusting the [WINDOW] settings directly.

Optional: Using a Table of Values

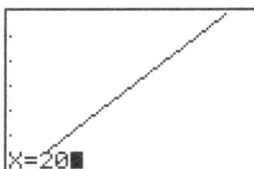

Press [2nd] [TABLE]. You should see a table of values containing a list of X-values and a corresponding list of **Y1** values that match your function **Y1=12X**. Use the arrow keys to navigate around this table. The example here shows how one student used the table to answer Question 1b.

To change the X-values in your table, press [2nd] [TBLSET]. **TblMin** is the starting X-value to start off the table. **∆Tbl** is the increment between any pair of consecutive X-values in the table. Experiment with the Auto and Ask settings to see if you can describe what they do.

Finding a Line of Best Fit and Using It for Prediction

Curve fitting means finding a function whose graph matches a data set closely. A curve that fits a data set well might not pass through every point in the scatter plot of the data. When you use a straight-line function to fit your data set, the best such line is called the "line of best fit."

This section contains instructions for using a graphing calculator to fit a line to a data set. The data set in this example is from the Jones family in *Sublette's Cutoff Revisited*.

Entering Your Data Set

To enter data, first press [STAT] [ENTER] to get to your screen of lists. You will put your data in lists L1 and L2. If there are any items in either list, put your cursor on the list name at the top of the list and press [CLEAR] [ENTER].

Use list L1 for the list of days and list L2 for the gallons of water. Enter all three data points into the lists. Use the arrow keys to navigate around the lists. Make sure the days and gallons match up correctly.

Graphing Your Data Set

To set up your calculator to graph your data set, press [2nd] [STAT PLOT]. You should get a screen like the one shown here.

You will use Plot 1. Press [ENTER] to display the Plot 1 screen. Make your Plot 1 screen match the one shown here by highlighting and entering the correct options. In the last line choose any mark you like.

Continued on next page

Before you graph, check a few things. First, go back to the [2nd] [STAT PLOT] screen and make sure all the other plots are turned off.

Also, press [Y=] and make sure all the functions are either deleted or turned off. At this point you don't want any extra graphs on your screen. Now press [GRAPH].

It's quite possible that you can't see any of your data set right now! You probably need to adjust your window to get a better view of your graph.

Adjusting Your Window to See Your Data Set

```
WINDOW
 Xmin=0
 Xmax=10
 Xscl=1
 Ymin=0
 Ymax=60
 Yscl=10
 Xres=1
```

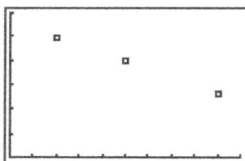

To view the range of your window, press [WINDOW]. Because the days in your data set range from 2 to 9 and the gallons range from 26 to 49, the screen shown here displays one reasonable window for this data set. **Xscl=1** instructs the graph to show a tick mark for every unit on the *x*-axis. **Yscl=10** instructs it to show a tick mark every ten units on the *y*-axis. Enter these window values from the screen here into your window.

Another way, to quickly find a good window for a statistical plot is to press [ZOOM], select **ZoomStat**, and press [ENTER].

To view your graph at any time, press [GRAPH].

Making a Guess for a Line of Best Fit

Make an informed guess at an equation of a line that you think will fit your points. Enter the equation into **Y1=** in the Y= edit screen.

Displaying the Function and the Data Set

To display both the function and the data set, press [GRAPH]. If your guess is far enough off, it might not show on the screen. You might need to use **Zoom Out** to search for the graph of your function.

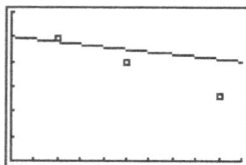

Continued on next page

Improving Your Line of Best Fit

You can probably improve on your first guess for a line of best fit.

Enter a new equation that is an improvement on your first line. If you like, leave your previous function in Y1 and put your second guess in Y2 so you can compare them. Continue improving your line of best fit until you are happy with it. Be patient; this could take you many tries.

Using Your Line of Best Fit to Make a Prediction

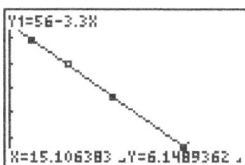

You can use your line of best fit to predict how much water the Jones family will have on Day 15. Use your calculator's TRACE feature, or its 2nd [TABLE] feature, or 2nd [CALC] 1: value, to find the point that matches Day 15. If you use TRACE, use the left or right arrow key to move the cursor along your best-fit line while watching the coordinates at the bottom of the screen. You may need to increase the size of your window.

Note: The TRACE feature moves the cursor along any graphed Y= functions or statistical plots. If you find you are tracing along the wrong graph, use the up or down arrow key to move the cursor to the correct function.

What does your line of best fit predict? Do your think your prediction is reasonable?

Now use your graphing calculator skills to complete the activity, *Sublette's Cutoff Revisited.*